The Universal Force

The Universal Force

Louis A. Girifalco

OXFORD
UNIVERSITY PRESS

Great Clarendon Street, Oxford OX2 6DP

Oxford University Press is a department of the University of Oxford.
It furthers the University's objective of excellence in research, scholarship,
and education by publishing worldwide in

Oxford New York

Auckland Cape Town Dar es Salaam Hong Kong Karachi
Kuala Lumpur Madrid Melbourne Mexico City Nairobi
New Delhi Shanghai Taipei Toronto

With offices in

Argentina Austria Brazil Chile Czech Republic France Greece
Guatemala Hungary Italy Japan Poland Portugal Singapore
South Korea Switzerland Thailand Turkey Ukraine Vietnam

Oxford is a registered trade mark of Oxford University Press
in the UK and in certain other countries

Published in the United States
by Oxford University Press Inc., New York

First published 2008

British Library Cataloguing in Publication Data
Data available

Library of Congress Cataloging in Publication Data
Data available

Typeset by Newgen Imaging Systems (P) Ltd., Chennai, India
Printed in Great Britain
on acid-free paper by
Biddles Ltd., www.biddles.co.uk

ISBN 978–0–19–922896–6

1 3 5 7 9 10 8 6 4 2

Let me tell you about Galileo, Newton, and Einstein, and about gravity, planetary motion, the Moon, and the stars. It's all wrapped up in one simple statement. Here it is:

The laws of nature are the same for everybody, everywhere.

Acknowledgements

I am deeply indebted to the following individuals for their comments and discussions: Hugh Van Horn, David Welch, Graig Welch, and Bill Bruehl. Among them, these individuals represent vast experience and great distinction in astronomy, theoretical solid state physics, art, and creative literature. Each of them carefully read my initial manuscript and offered invaluable criticism and advice. They were generous with their time, and their contributions were acutely perceptive. I am also indebted to Nicholas Cordero, who painstakingly went through all the scientific content to verify its accuracy and in so doing corrected some errors in my presentation. My manuscript was much improved by them and I hereby express my gratitude.

Contents

Personal prologue

Walt Whitman's "Leaves of Grass" had a special attraction for me. It was not the objective content or even the subjective interpretations I might impose on it that held any special meaning. It was the music. The way those words rolled on, the rhythms and sounds that flowed from here to forever, capturing the movement of time, made a connection with something basic and primitive. That compound of desire, awe, and mystery that all young people feel was crystallized for me at a purely non-verbal level by Whitman, and I labeled him great.

But later, when I saw the phase rule and remembered his poem "When I Heard the Learned Astronomer", I felt sorry that Whitman could only touch a part of the mystic sense.

I had learned the phase rule in a physical chemistry course at Rutgers. It is a simple thing connecting the simplest properties of ordinary matter, and arises from just counting the number of certain thermodynamic equations, but it governs the conditions under which all the different kinds and states of matter could coexist.[1] In a purely intellectual way I knew that its power was immense, but I did not really appreciate that until I was walking along Raritan Road in Linden one winter night about a month or so after I had completed the course.

It was not unusual for me to walk at night. There were many days when I got home from school after six and didn't start to study

[1] The phase rule contains a few simple ideas. The first is that of *phase*. This is just the state in which a material exists. Water, for example, can be in any of three phases; solid, liquid, or gas. The notion of component is even simpler. Water consists of one kind of molecule, so it has one component. Brass consists of copper and zinc, so it has two components. *Degrees of freedom* define the environment; temperature and pressure are degrees of freedom. The phase rule states that the number of phases plus the number of degrees of freedom is two more than the number of components. We memorized it using the phrase "police force equals chief plus two".

and do homework until after eight. By midnight or one o'clock I was losing the ability to concentrate, the ideas and images were starting to race around without control, and there was no more order in my mental world. I was tired enough that it made no sense to keep working, but it was not possible to get to sleep, so I went for a walk. After a mile or two the turmoil in my head would quiet down and I could go home and go to bed. We lived on the sparsely populated outskirts of town, more rural than suburban, and the walk was usually quiet, calming, and pleasant.

This night was really extraordinary. A combination of snow, sleet, and rain had left everything coated with ice. It was late enough that nobody was on the streets; all was silent and no lights were on. But the sky was marvelously clear and bright. In those days the stars over Linden were not merely visible, they were brilliant. Fence posts, leaves, and tree branches were all covered with a hard, transparent layer that captured the light from the sky and sent it back in all directions. The trees were particularly striking. They were made of glass, full of small shining point sources, and were the dominant factor that made everything seem magical and strange. I was not yet twenty years old and could still respond to the incredible beauty of such a scene instantaneously, with wonder and delight. It all sounds a bit callow and banal now, encumbered as I am with decades of ordinary reality, but at the time it was an experience of Nature's poetry, more intense than Whitman's music.

The temperature must have been quite close to the melting point because there were places where ice had melted and was in contact with water. And there it was: ice, liquid water, air, water vapor, wood, telephone wires, sidewalks, and stones, coming together to form a stunning visual harmony in which the temperature, pressure, and material parts were all connected by the phase rule. The knowledge of that simple mathematical relationship and its embodiment in that night of crystal multiplied its beauty. The beauty existed because of the mathematics, and the mathematics organized the beauty, and in my arrogance I felt that I was in touch with something profound and fundamental. The reason I had been drawn to study science lay before me, powerful and insistent. It was an attraction born of mysticism; I could not believe that the world was just there, prosaic and ordinary, without something marvelous at its core. I wanted to come face to face with the awesome and wonderful, to immerse myself in it, and above all to see its structure and foundations. I wanted to know IT, the secret of everything, to understand, and therefore command, the elemental forces of nature.

I now know that this is a common motivation for studying science. Newton's locked chest of papers shows that, at bottom, he was the ultimate mystic, and surely the desire to come to grips with fundamental mysteries must be a strong part of anyone who takes science seriously. We want to know the deepest secrets of nature, and we do not believe that emotions, revelations, or magic can help us. Science is the only method of getting at that hidden treasure that is not completely laughable.

But poor Whitman couldn't see this. He was clearly confused and bored by a lecture on astronomy. He thought it was cold and sterile and had nothing to do with beauty or the mystic sense. Before the lecture was over, he went outside and "looked up in perfect silence at the stars". He did not understand that an appreciation of gravitational forces, the laws of motion, and theories of stellar evolution enhances the sense of wonder manyfold and brings one *inside* the mysterious, as a participant, not just an observer. I learned later that I was not the only scientist that had loved Whitman and was then disillusioned by that terrible poem.

There are a great many people like Whitman. Followers or creators of literature and art often see science as something inhuman or even anti-human that has nothing to say about the really important aspects of the human condition and is actually hostile to our best aspirations. Perhaps they see science as the origin of technologies that are responsible for great evils from alienation to the H-bomb; perhaps they dislike scientific analysis because it breaks the world into parts to be studied instead of a whole to be simply experienced; perhaps they will not accept the restrictions on the meaning of reality imposed by science because such restrictions limit the freedom of the imagination; perhaps they don't want a competitive priesthood whose practical power is much greater than theirs. And perhaps they just don't understand the language of science, whose mastery requires a kind of intellectual discipline and rigor that is often alien to the artistic mind.

At any rate, they are missing a lot and are always in danger of being trapped by the unreal.

Yet, there are many who are not at all like Whitman. Most of them are in the same category in which I define myself, although offering insights that are very different than those of science, but yet exciting and even compelling.

I hereby invite them all: those who are like Whitman and those who are not, scientists and non-scientists, to explore the spirit and excitement of science. They will see that we all have a great deal in common.

The close connection between science, beauty, and the mysterious can be found everywhere. The most striking rainbow I ever saw was on a summer day a few hours before sunset at the Jersey shore. It was a perfect double bow, clear and bright with vivid colors, and it was very beautiful. Its beauty was magnified by my knowledge of the double refraction in water droplets that produced the colors. The secret of the rainbow is in the twofold bending of light in tiny spheres of water. Certainly a non-scientist can appreciate a rainbow, but I think scientific insight makes it all the more wonderful.

And then there is the Verrazano Narrows Bridge.

That astonishing bridge appeared through a light haze, just as I was taking a slight curve on the Belt Parkway. All suspension bridges are inherently beautiful, but there is nothing comparable to this. Those huge towers soaring to the sky, strung with yard-thick steel cables, holding up a 4260-foot span with an exquisitely tuned balance of forces, curving down almost to traffic level and then back, with magnificently proportioned curves, delicate and ethereal, created a wordless poem, a work of art comparable to any symphony.

The mist was thin, so the visibility immediately ahead was excellent, but the bridge was still far away and light could not penetrate the distance; and the mist was low, covering the water like a light translucent blanket turned hazy white by the early sunlight streaming from above and scattered by innumerable microscopic droplets. It was a vision from a dream, seeming to defy natural law, a great, organized symmetry of steel lace, graceful and light, floating on an insubstantial vapor with nothing holding it up, stretching to the invisible far shore, narrowing and disappearing as if chasing infinity. Its beauty was stunning and it moved me as only certain manifestations of the awesome could move me.

It was gravity that gave the bridge this air of unreality, isolated from all else, and dramatically accented by those great loops of steel whose conformations were dictated by engineering necessity. Its beauty was its own but gravity was the essence that made it look so spectacular.

All pervasive and universal, the force of gravity dominates every moment of our existence and permeates our very bones. It is so ubiquitous that we deal with it automatically, by reflex, as we respond to the atmosphere by breathing, and yet we are always aware of its power. It seems to be an unfathomable mystery. Much of our wishes, wizardry, and dreams are tied to our desire to overcome this thing that glues us to the ground. Technological advances, from Icarus to the space shuttle, only serve to deepen the mystery and verify the awful fact of gravitational force.

That great bridge, on that enormous vaporous bed, seemed to float without support, apart from the rest of the world and disconnected from reality.

The phase rule, rainbows, and gravity are but three examples of the strange poetry of science, and gravity is the strangest, the most universal and the most compelling.

Preface

The dramatic manifestations of nature—storms, lightning, volcanoes, rainbows, sunsets and stars—all cry out for understanding, for exposure of their inner mysteries. We recognize that these awesome displays hide great secrets of profound meaning, secrets even more astounding than the displays themselves. They are powerful, undeniable facts that would expose the inner workings of nature and say something about the significance of the world, if only we understood them.

Thoughtful contemplation leads to the conclusion that ordinary processes and objects are no less mysterious and demand no less an explanation. A breeze or a flowing stream is as much of a wonder as a hurricane or a tornado; a stone is as much a mystery of existence as a star or a galaxy; the colors in a small oil film are as beautiful and demand as much explanation as a rainbow or a sunset. The ordinary is just as much a puzzle as the spectacular.

The complexities of nature, whether common or rare, must be explained in terms of more basic things, and the search for these deeper simplicities has been continual since the beginnings of intelligent consciousness. And always the greatest seekers have come to focus on the nature of motion and of gravity, sensing that an understanding of these simple phenomena would provide the key to unlock the secrets of the world.

Motion is just the movement of material objects from one place to another, and it intimately involves space and time, those two inescapable concomitants of human awareness, so bound up with the human psyche. And gravity is the predominant presence of our daily experience. Pervasive and limiting, it keeps us glued to the Earth like impaled insects, and allows us to rise above it, in airplanes or rocket ships, only with the expenditure of great stores of energy. Our dreams of freedom and magic are dreams of flying and levitation, in which gravity is no longer in control and our movements are what we wish them to be.

Gravity demands to be understood, and I hereby present the result of my own search for understanding, a search that consisted of trying to learn from the achievements of those great scientists who have struggled with the seemingly simple facts and managed to extract some truth about the nature of gravity, its origins, and its effects. Gravity is intimately tied up with motion, and therefore with time and space, so to look at gravity is to look at the deepest aspects of nature.

The story is not completely linear. There is a rough history of gravity that follows an approximate time sequence, but it is strewn with ideas and events that do not fit into a neat temporal frame. The reason for this is that gravity has surprising connections with other natural phenomena. Light, electricity, and magnetism, and the fundamental nature of time and space, as well as motion, are intimately bound up with gravity. Our intuition is right. Understanding gravity provides great insight into how the world works.

I have known a great many highly intelligent people with special talents and expertise in a variety of non-scientific fields, ranging from musical composition and poetry to history and Renaissance literature. The best of them share common intellectual characteristics. They take little for granted; they use careful logical reasoning; they search for patterns and connections; they sometimes gain insights without knowing how. And they are all enthralled with the inherent beauty of what they learn and create. Many of them, however, regard the closely reasoned, stark nature of abstract scientific thought as a barrier, rather than an aid, to understanding. And so many are put off by even the simplest mathematics. They are representative of the "non-scientific" intelligent public.

Scientists, however, have been trained to think in the language of mathematics and abstractions. They enjoy logical puzzles and get as much pleasure in understanding scientific issues as any non-scientist gets from successfully absorbing great literature. Numbers are their friends, just as words are friends for the literati.

I want to reach both the scientific and non-scientific types. It must be possible, because both are after a deeper understanding of the world. Yet there is this thing about mathematics, and to deal with it I have adopted the following procedure. There is no formal mathematics in this book. Much of the important science of gravitation can be presented and appreciated without it, so I have presented even the quantitative scientific content descriptively. I have found this to enhance my own understanding, because mathematics is a shorthand that can sometimes condense and mask physical reality, even as it lays bare the basic relations in nature. But I could not resist putting in two equations: Newton's equation for the law of universal gravitation and Einstein's equation for the equivalence of mass and energy—no mathematical manipulations, just these two equations. They are just too elegant and important to ignore, and I wanted to make a point. They are so simple that they take up very little room on the paper and they *look* simple. Each involves only a few symbols and yet each describes nature in its deepest sense and each has an almost infinite number of important consequences. And I do not think they are any imposition on the non-mathematical reader. Surely every educated person by now knows the mass–energy equation, and the inverse square equation is very easy to understand. But if the equations annoy you, just ignore them.

Galileo's works are still among the finest examples of science writing for the "intelligent layman". He presented practically all his work in a form easily accessible to non-specialists with a minimum of mathematics, clearly and cogently. I believe this can be done for any science and I am trying my hand at it here.

Some of my own unhappy experiences with learning mathematics convinced me that many people who "can't do math" or "don't like math" are simply victims of bad teaching in elementary and high school. The curriculum usually focuses on methods and problem solving. The emphasis is all on process and getting a "right" answer. Very little attention is given to what mathematics really *is*, or what mathematical objects and manipulations actually *mean*, or to the insight that it involves only simple logic. The few attempts to address these issues are often ridiculously wrong headed, as in the past vogue of trying to teach set theory in grade school. This is a pity, because it is easy to teach math properly. There is no place for such a digression here, but readers should rest assured that there is nothing beyond their abilities in this book.

Of course, there is no way to lift all burdens from the reader. Nature has simple foundations, but its various, detailed expressions can be complex, so some concentrated effort and careful thinking is required. The piano can be mastered only by the hard work of practice; the joy of great literature can only come from a careful reading of good books. Similarly, the pleasures of science can only be experienced by those willing to expend some effort.

The most important point for the non-scientist is not the lack of mathematical expertise, but the requirements of abstraction. In thinking about the law of falling bodies, as exemplified by Newton's legendary apple, the imagination must be restricted to a consideration of time, distance, mass, and gravitational force. Everything else gets in the way. It doesn't matter if the apple is ripe or not, or if it is a Granny Smith or a Red Delicious. It doesn't matter if it is bruised or if the tree it came from is young or old, or if the tree was in a neighbor's garden. All of these factors might be important for a poem or a love story or a neighborhood vendetta, but they have nothing to do with the physics of falling bodies, which abstracts only time, distance, mass, and gravity from the great many characteristics surrounding a falling apple. The abstraction even ignores air resistance, temperature, and the nature of the ground the apple falls on. They are just not relevant. Everyone will agree that irrelevancies must be ignored, but when highly intelligent people say they do not like mathematics, what they often object to is a chain of thought in which everything but a few key concepts have been thrown away. Then only the data and logic remain. The intense concentration needed takes work.

The artistic mind wants to explore all possibilities: the relationship among *all* properties of the apple, including its possible role in human expressions of love and hate, and the example it gives us of birth, maturation, and death. Compounding the problem of abstraction is that of definition. Scientific thinking requires that we be careful of what we are talking about, so we must be precise with the meanings of the words we use. Here is a simple example. In normal usage, the word "power" has a number of meanings and connotations. It could refer to the influence of a politician, to the authority assigned to an executive, to the strength of a rioting mob, or to the character of a torrential storm. The word has a variety of meanings and connotations depending on the context in which it is used. In physics, however, it has a specific and unchanging definition. It is the rate of doing work. And in this definition, the word "work" has a carefully specified meaning that has nothing to do with the many ways the word is used in everyday speech. If the reader can keep in mind the importance of abstraction and of the precise use of defined terms, a lot of the problems the non-scientist has with scientific subjects will vanish. For any reader of science, no matter how elementary or how advanced, here is an essential rule. *Know what each word means.* If you have trouble with any scientific explanation in this book, the first thing to do is make sure you know the correct meaning of the words. Precise vocabulary is everything. The rich ambiguity of language is often important for the artistic merit of literature, but in science, precision of definition is essential.

Also, while everyone knows how to think logically, it often takes a good deal of concentration to correctly follow a logical argument. Even if the vocabulary is thoroughly understood, and even if there is an intense desire to understand, the mind can wander or get tired when trying to follow a closely reasoned argument. The only defense against this is to remember that the logic is just the logic we are all used to, and to try again when we get off the track. Those who aspire to read this book can certainly summon the discipline and effort to do this. There is nothing that is beyond the abilities or knowledge of an intelligent human being.

The scientific enterprise is an art of its own kind. Of all the arts, it is closest to musical composition, which requires the same kind of previously unimagined insights and probing beyond limits, while strict attention is paid to those limits; the same kind of search for, and construction of, tight relationships; and the same kind of evolution and maturation of ideas. And it yields the same dividends of beauty of structure and content and even the thrill of the "unexpected inevitability".

Scientists are fully aware of the beauty and mystery in nature. At bottom, science is mysticism, in the sense that it is driven by the need to probe into the greatest of mysteries, even though we know that the search cannot ever be completely successful, but also that the scientific enterprise holds the only hope of getting at true knowledge about the physical world. And so we study, do experiments, calculate, and theorize. We enjoy unraveling puzzles about nature; we enjoy the artistic aspects of good science; and, coupled with these intellectual delights, there is the immense emotional pleasure of looking into the most mysterious and the most awesome.

Many books have been written to explain the results of science to non-scientists and, on the whole, they constitute an excellent literature that does a good job of making science comprehensible to the legendary "intelligent layman". Anyone willing to devote some time and intellectual effort can find good popularizations devoted to almost any scientific subject, from molecular biology to cosmology. And often very little scientific background is required, because scientific thinking is just like any other sound thinking. It depends only on the consideration of facts and on the well-known processes of logic, so presentations of even the most esoteric science can be made understandable. It is worthwhile repeating that often the only barrier to understanding is vocabulary. It is essential to clearly define the meaning of the words we use and to note that these meanings are not always the same as those in common, non-scientific usage. I agree with the observation of the great physical chemist who said, "No theory is worth anything unless it can be explained to a college sophomore". But the meaning of the words used must be clear.

I want to do more than explain the facts and theory of gravitation. Science has its roots in a special kind of mysticism, an insistent desire to know, to unravel mystery, to see what the physical world really is and how it works. It has nothing to do with what is usually called mystery. For science there are two equivalent statements: the first is that there is no mystery. The second is that *everything* is a mystery. And I think the best way to approach this "mysticism of the real" is to follow its development through a look at the men who led the struggle to understand.

There is no belief in magic, or the supernatural, and above all, there is no rejection of logic or reason in science. It is based on the assumption that we can learn what is to be known about nature from the evidence of our senses and the application of sound logic. The kinds of revelations arising from altered states of consciousness, whether induced by drugs, fasting, or meditation have no place here. They are private and personal, cannot be communicated to others, and say nothing about objective reality.

Science is attractive because of what it reveals about nature and its intrinsic beauty. There is an unparalleled thrill in penetrating the unknown and in finding out how the real world really works. The most remarkable aspect of science is that, starting with just the evidence of our senses and using ordinary, mundane logic, we arrive at the most amazing conclusions about the structure and workings of the world. This is particularly true of gravitation. It holds the universe together and it is the force that we must continually overcome. The galaxies, the stellar systems within them, the planetary system, and the Earth itself are held together by gravitational force. It is responsible for the formation of stars and planets, yields strange objects like neutron stars and black holes, and is the most constant aspect of our daily lives. It is intimately tied up with the deepest structure of space and the flow of time, the nature of light, and the mysteries of electricity and magnetism.

Gravity is the ultimate and universal force. Every bit of matter, and in fact every bit of energy, exerts a gravitational force on every other bit of matter or energy. There is no exception. And when particles are so far apart that no other forces can act between them, when positive and negative charges neutralize each other so that electric or magnetic forces are cancelled, or when they are so close together that they lose their identity, gravity is still there. And it reaches out forever into space, putting everything that exists under its influence. Gravity is involved, essentially and inescapably, in the beginning of the universe and determines how it will end. The formation and evolution of galaxies, the stars, the Earth, the Moon, and the planets are all controlled by gravity. It is a truly remarkable phenomenon.

I invite you to share its wonders.

1

The seeker

In a miraculous life, it was a miracle that he was alive at all. He was born on Christmas day in 1642 in a small English town called Woolsthorpe. He weighed less than two pounds, so small that he could fit "into a quart pot". It was a time when infant mortality was high, nothing worthwhile was known about medicine, and both prenatal and postnatal care were ridiculously crude and wrong-headed, so that many babies, and practically all premature or seriously underweight babies, died quickly after birth. In fact, Isaac Newton was not expected to live, and his survival was a surprise.

He had other misfortunes. His father died three months before he was born, and while his mother waited a respectable three years to remarry, when she did so she went to live with her new husband and abandoned Isaac to the care of his grandparents. She returned to him after her second husband died, when Isaac was about ten years old, with the hope of having him ultimately take over the family farm. While he was well cared for with respect to the necessities of life, he saw little affection or love in his formative years, a lack which was not relieved by his mother's return. She sent him off to King's School, far enough away that he had to room with an apothecary. He grew to be a silent, secretive, testy individual with few, if any, close human relationships, spending much of his time alone, reading, dreaming and exercising his remarkable manual dexterity by building working models of everything from windmills and sundials to water wheels and clocks.

A lonely, solitary child, he grew to be a lonely, solitary man. His lifelong habit of keeping extensive notes on everything he thought and read started while a schoolboy, and they show that he was thoroughly unhappy, not knowing "what employment he was fit for", unsure of what to do with his life, and sometimes even contemplated suicide. His mother, on the other hand, had definite ideas about his future and brought him home from school when he was sixteen years old to take his place on the farm.

His mental powers were prodigious, and although he did not discuss his inner life with anyone and kept much of it secret, a few perspicacious individuals saw him for what he was. His schoolmaster and uncle were instrumental in saving him from the life of a farmer, for which he was most unsuited, and, with his mother's reluctant agreement, having him sent to Cambridge at the age of nineteen.

Newton had been unregenerate in neglecting his duties in favor of reading, model making, and solving math problems, so it became obvious he would never be of any use on the land. His mother's second husband left a substantial estate and Hannah Newton-Smith had more than enough money to pay for Newton's schooling and support. However, she gave him only a small allowance, and Newton paid his own way by being a servant to other students.

Many of the remarkable advances of the sixteenth and seventeenth centuries took place in continental Europe, especially in Italy, which had a vibrant intellectual climate nurtured by the Renaissance, until the Catholic Church retarded the growth of Italian science by condemning Galileo. The great scientific movements then grew in other parts of Europe, especially England. But the academic institutions did little to support the most innovative science, and much of it took place only because of the efforts of individuals. Cambridge in particular was a backwater, not only geographically, but also intellectually. It had absorbed none of the fruits of the great intellectual ferment that was sweeping away the old order. Its curriculum concentrated on Latin and Greek and the ancient Aristotelian corpus. Newton had a raging passion for learning, absorbing everything Cambridge offered, but critically, with a fine eye for the unproven, the illogical, and the useless. He sought knowledge everywhere and taught himself the new science by hunting down and devouring the books and manuscripts in which it was described.

Newton received his degree in 1665 and was soon elected a Fellow of Trinity College. That same year, Cambridge was closed because of the plague, which was first recognized when two sailors were found dead with the characteristic swollen purple buboes of the Black Death. Previous instances of the plague pandemic were so devastating for all of Europe that anyone able to escape crowds did so. While not as virulent as that of the fourteenth century, the plague of 1665 was bad enough, some estimates stating that as much as one sixth of the population of London died in one year before the disease ran its course.

So, along with most others who could, Newton left Cambridge to be away from crowds and spent the next two years on the farm at Woolsthorpe. They were by far the most amazing and productive two years spent by any scientist before or since, matched in intensity and accomplishment only by Newton himself twenty years later during the several years before completion of the *Principia*, the monumental work that created modern mechanics and correctly derived the properties of the solar system. The only other period that came close in creativity was the "miraculous year" of 1905, in which Einstein worked out the theories of special relativity, the photoelectric effect, and Brownian motion, each of which initiated a critically important branch of science.

Until the seventeenth century, geometry was the most important mathematics. Supplemented by trigonometry, it was the language of kinematics[2] and astronomy and provided the best paradigm for rational truth obtained by deductive reasoning from

[2] In physics, kinematics is the science of the motion of matter having nothing to do with forces. That is, it studies the position, acceleration, and paths of motion. Dynamics is the study of motion that includes the forces acting on the body as well as its position, velocity, and acceleration.

a few incontrovertible assumptions. Algebra, which had been developed and augmented to a considerable extent since it was adopted from Arabian scholars, was also in use but possessed neither the range nor the authority of geometry. Descartes changed the entire landscape of mathematics with the creation of analytic geometry, which was the true beginning of modern mathematics and made the calculus possible. Newton had absorbed all existing mathematics, including Descartes' work, and had gone beyond it. During the plague years, alone on a farm, he became the greatest mathematician in the world and took a giant leap beyond analytic geometry by inventing the calculus, which he called the "method of fluxions". The term "fluxions" is revealing. Calculus deals with the changes of quantities relative to each other. In a square, for example, the area is the square of the length of one of its sides; doubling that length increases the area by four times, and the ratio of a very small change in area to a very small change in length is twice the length of its sides. The ratio of small changes in *any* two related quantities can be calculated. But Newton thought of each of the two quantities as changing in time, and a fluxion is the rate of change of the quantity with time. It was a kind of velocity and is an example of the derivative, the fundamental concept of the differential calculus, which is easily understood by considering the motion of a single particle along a straight line. Let's measure the position of the particle at two different times so that we have the distance traveled over some time interval. Then the average velocity is just the ratio of the distance traveled to the measured time interval. The value of the velocity might be changing during this time interval, and it would be desirable to track this change by having the instantaneous velocity for each instant of time. This can be approximated by making the time interval small so that the distance the particle travels is small. Then the ratio of the distance to the time interval gives an average velocity during a small time interval and approximates the velocity at a point in the path of the particle that is in the middle of the distance traveled. If the time interval is made still smaller, the distance traveled is still smaller and the distance/time ratio is a yet better approximation to the instantaneous velocity. If the increments of time and distance are made ever smaller, the ratio approaches a limiting value called the derivative of position with respect to time. It is just the instantaneous velocity of the particle.

The generalization to any pair of quantities that depend on each other is straightforward. The volume of a gas at constant temperature, for example, depends on its pressure, and the derivative of the volume with respect to pressure is the ratio of an infinitesimally small volume change to the corresponding infinitesimally small pressure change. The mathematical structure erected on this simple idea is very powerful and permits the description of natural laws of great generality.

Newton's fluxions are time derivatives, and the general derivative, as we know it, is the ratio of fluxions. The roots of the calculus, therefore, were in the analysis of motion, its most important application to physics, and Newton was then adopting his concept of time as a uniform flow independent of anything else.

Monumental though it was, calculus was not Newton's only accomplishment during those two years at Woolsthorpe. He had studied all Descartes' other writings, as well as his mathematics, and was familiar with all current theories of motion, gravitational action, cosmology, and optics. His experimental work with prisms showed him that white light was a mixture of colors and resulted in his great work *Opticks* published many years later. As if this were not enough, the foundations of his theory of universal gravitation and the laws of mechanics were also laid down during those extraordinary plague years.

To this day, we must stand in awe of what he accomplished.

Newton's personality was as unusual and extreme as his intellectual powers, with many contradictory elements. He was a loner, a self-medicating hypochondriac, arrogant, secretive, and suspicious, always reluctant to reveal the nature or results of his work, yet so jealous of recognition that he was led into long public arguments with the most distinguished of his contemporaries on questions of priorities.

The triple symbolism of having been born on Christmas day, in the same year that Galileo died,[3] and surviving a most dangerous infancy was consistent with his self-image. He knew that his intellectual gifts were far beyond those of others, and he thought of himself as a unique historical figure placed on Earth by God to penetrate the deepest secrets of the universe. His greatest desire was to be left alone to fulfill his destiny. At the same time, he craved recognition and doted on the public admiration arising from his continually growing reputation as the greatest natural philosopher of his time. In his later years, he even accepted a government appointment, a move that no one knowing his earlier life would have predicted.

Issac Newton has become the symbol of that mechanistic view of the world that has been blamed for the decline of religion, the rise of crass materialism, and the advance of technological growth. His universe is believed to be a clockwork universe that ignores human values and has no place for spirituality. He is thought of as the epitome of the cold, calculating scientist, a rationalist who discarded everything except experiment and logic in seeking knowledge. To many, he is therefore the most important progenitor of the modern search for knowledge and its benefits. To others, who view scientific and technological progress as the source of much more harm than good, and as antithetical to the aspirations of the human spirit, Newtonian mechanism is the enemy of the arts, of the mystic sense and a threat to the very survival of humanity.

This view is a total travesty of the man and his work. I can do no better than quote John Maynard Keynes, who, on the basis of Newton's secret, unpublished manuscripts, said: "Newton was not the first of the age of reason. He was the last of the magicians, the

[3] This happy aid to memory is not quite accurate. The Gregorian calendar was not adopted in England until 1752, but it was in effect in Italy when Galileo was born. According to the Gregorian calendar, Newton would have been born in 1643.

last of the Babylonians and Sumerians, the last great mind which looked out on the visible and intellectual world with the same eyes as those who began to build our intellectual inheritance rather less than 10,000 years ago. ...the last wonder-child to whom the Magi could do sincere and appropriate homage". Keynes is justly famous for his landmark works in economics. It is less well known that he was first an excellent mathematician and wrote an outstanding book on probability. He was a multifaceted genius. His remarks about Newton were written for the tercentenary celebration of Newton's birth[4] and were based on his reading of materials in Newton's locked box, which showed that Newton spent a great deal of energy on pursuits we can only call "magic".

Descartes, not Newton, was the true believer in a clockwork universe, needing God only as a First Cause to get it going, from which point it proceeded on its own according to rigid mechanical laws. The mechanistic philosophy was completely alien to Newton's world-view. He believed that God *continually* acted to maintain the processes of nature, and this was one of his disagreements with the views of Leibniz. Newton knew that there were slight irregularities in the orbits of heavenly bodies that would eventually radically change them, and he described this as an "unwinding of the world", which would then need divine intervention to restore it. He concluded that Providence was always present to make nature go. Leibniz could not accept this because it implied that God did not create a perfect natural system, and he could not accept such an imperfection. He therefore held that nature was perfect and needed no further fine-tuning. In this, he agreed with Descartes.[5]

[4] Keynes died before the celebration took place and his lecture was read by his brother.

[5] There is a contradiction in Leibniz's thinking here. In another context, Leibniz approaches the problem of God creating a world in which evil exists by saying that God did not create a perfect world because then it would be merely an extension of Him and not a new creation. This is similar to some modern theological positions.

We must understand that Newton's purpose was not just to find the laws of mechanics, or the law of gravitation, or to understand the motion of the planets, Moon, and stars, or to learn how tides were generated, or to study light and color, or to create a world of new mathematics, or to perfect a new kind of telescope, or to create mathematical physics. He did all this and more, but his ambition went far beyond these stunning achievements.

Newton wanted to *know*. How was the world made? How did its parts relate to each other? And why? He believed in the unity of the world, of both its material and its spiritual parts. The motion of the planets and of terrestrial objects; the colors of the rainbow; the flow of rivers and the motion of tides; the blue of the sky and the green of grass; wind and weather; condensation, vaporization, melting, and freezing, chemical changes; the forms of plants and animals: all of it was part of a single fundamental truth. And it was not independent of God. On the contrary, it was from God and a part of God, and Newton's craving for knowledge extended to the nature of the Deity itself. He was convinced there was a unity and he was convinced he was destined to find it. To this end, he studied theology and its history, the Bible, and alchemy, as well as natural philosophy.

Newton's belief in a fundamental unity was a problem professionally because he was a fellow of Trinity College and required to confess the true Anglican faith.[6] Privately, he adopted the Arian heresy. He could not accept the idea of a Triune God and took God to be One, Single, and Undivided with no parts or multiplicity, beyond time and space, the origin of all and the cause of all.

His studies in theology were no less thorough than those in natural philosophy and alchemy. His knowledge of the Scriptures was complete and he was an originator of the critical study of the Bible. He was also a student of the writings of the early Church

[6] A provision of the Cambridge Lucasian professorship was that the holder would not be active in the Church. Newton successfully used this provision to avoid taking Holy Orders.

Fathers and absorbed everything written by the religious scholars. From a long and intense historical analysis of the Bible, and a study of the evolution of Christian doctrine, Newton concluded that the prevalent idea of the Trinity was added to early belief by the Council of Nicea and that Arius, who contended that Christ and God were not identical, was right. The Trinity was an illogical perversion of the original, true Christianity. So, on the basis of logic and historical scholarship, Newton was a Unitarian. He was the ultimate reductionist, believing that all knowledge of all types could be reduced to a single origin and a single, all-encompassing system. He was deeply religious, taking care to seriously answer any suggestion that his work made God unnecessary. He restricted mechanistic explanations to the physical world and added a discourse on the role of God in the second edition of his *Principia*. "This most beautiful system of the Sun, planets and comets, could only proceed from the counsel and dominion of an intelligent and powerful Being".

It is instructive to recall the origin of the term "natural philosophy". To the ancient Greeks, knowledge had two methodologies. One was called dialectic philosophy, whose purpose was to pursue knowledge using pure reason to obtain a complete and systematic world system, encompassing all and explaining all. Parmenides and Plato were the prime examples of this approach. The other method was that of natural philosophy, in which knowledge, including philosophy, was to be built by reason, but using observational data as its basis. Aristotle and Archimedes were the most notable proponents of natural philosophy. In essence, one method sought knowledge by talking about it, while the other actually examined the world to find facts.

The giants of the one hundred years leading to Newton had done much to separate out science as a distinct study, but the general tenor of thought still held it to be subsidiary to the grand philosophical pursuit. This concept was supported by theology and was surely part of the ambience that drove Newton to seek a unified knowledge. It was part of Newton's genius that he was able to pursue mathematics and science separately with rigorous adherence to its own methods, apart from theology and religion, while deeply believing it was part of the whole. Yet, the separation was in the air and, ironically, Newton's huge successes completed it.

He pursued every conceivable path to knowledge, the scientific path being but one. He read all the ancient works he could find, including those on alchemy and sorcery; he carefully studied the Bible, which he believed literally, and made calculations of the sequence of Biblical events from astronomical data. He spent untold hours at the furnace of his laboratory on alchemical studies, hoping to find the secret behind it all. In fact, later analysis showed that he had a lot of mercury in his hair, which would be expected because mercury had a special role in alchemical experiments.

Most of his life was spent on studying things other than physics and mathematics. Religion, mysticism, prophecy, history, the Bible, and alchemy commanded his attention for much longer times than did mathematics, physics, or astronomy, and the volume of his writings on these far exceeded that on scientific subjects. He thought it was all part of a single, coherent system of truth. He studied natural philosophy to see the Hand of God in the physical world. He studied the Bible and theology to understand God's Will; he studied ancient works, including the Greeks and beyond, because he thought they had a knowledge that was corrupted and lost by succeeding generations; and he studied alchemy because he believed that matter was fundamentally of one kind which could be transformed into different manifestations by an alchemical "vital agent". He tried to find this vital agent, which was supposed to be responsible not only for the usual chemical changes but also, and especially, for life processes such as digestion and growth. Newton could not believe that these were purely mechanistic and held that life processes required the vitalism that was basic to alchemy. He believed that atoms, acting of themselves mechanically, could not produce the endless variety of forms, particularly of living things, that were evident in nature. Ultimately, God was the cause of these and Newton sought the vital agent as the means through which God acted. In terms of the modern debates, Newton held to the concept of "intelligent design" and, like almost everyone at that time, he was a creationist.

Alchemy absorbed an enormous amount of his time and energy, in experimentation as well as in studies of the work of alchemists from ancient times to his contemporaries. He constructed a laboratory containing the latest equipment, including two furnaces, appropriate glassware, and a precision balance. Although the balance had been used in alchemy for decades, many alchemists generally paid little attention to the precise amounts of material used in their experiments. Newton, however, applied the same intense, quantitative precision to alchemy as to his work in physics.

There were two aspects to alchemy. One was centered on the Philosopher's Stone, which could be used to change lead into gold, and the Elixir of Life, which could confer immortality. Its practitioners were motivated by greed and often used their esoteric knowledge to cheat gullible nobles by promising what they could not deliver. The second aspect was very different. While not denying the possibility of a Philosopher's Stone or an Elixir of Life, it viewed alchemy as a legitimate method of discerning the works of God and had a spiritual basis. In fact, the seven steps of one basic alchemical process were thought to be a microcosm of the Creation. In this mode, alchemy was an important part of the philosophical life of intellectuals both in England and on the Continent. Of course, they scorned the cheats and charlatans who called themselves alchemists but whose only interest was private gain. The true alchemists were seekers of God's truth. They were a secretive bunch because they believed that alchemical knowledge in the hands of common people was dangerous, and all their writings were in allegorical form, often making no sense at all to a modern reader. It required years of preparation and study to become an alchemist and to understand the special language. Another reason for secrecy was that the authorities often frowned on alchemy and sometimes even outlawed it because they understood how the successful transmutation of lead into gold would upset the social and political order.

Newton met with other alchemists, read all the important alchemical literature, and performed many experiments, hoping that, in the study of material changes, he would find what was true and immutable.

Yet only his scientific work had any lasting value. His other efforts were the subjects of reams of notes, analyses, and speculations but yielded only a few trivial historical and religious publications. Nothing like the monumental *Principia* or the fascinating *Opticks* ever came of these metaphysical studies. It was his scientific and mathematical work that made him famous and transformed the world forever. And the sum total of the time spent through his lifetime on such work was not more than ten years. It is impossible to imagine how much greater and more far-reaching his accomplishments would have been if he had concentrated only on science and mathematics.

How could he ever have been a scientist of even average abilities, let alone the greatest who ever lived? Mysticism, in its generally accepted sense, admits of experience and causes that are beyond empirical reality, the universe being an ineffable entity beyond trivial scientific understanding. Its essential methodology consists of meditation and revelation, not scientific analysis and experiment.

Newton was not a mystic in this sense; his search for truth in non-scientific areas had, by definition, to be based on non-scientific methods, but they were completely rational. Rather than on an experimental and observational knowledge of the physical world, he had to rely on historical accounts, the meaning of ancient linguistic usage, decoding of cryptic writings, religious dogma, and the statements of those who claimed revelation as the source of knowledge. He approached this with the same industry and penetrating logic so evident in his scientific work. In light of this, his adoption of a method for studying nature, in which only experiment and reason count, may seem surprising. But Newton's mind was all of a piece, and he clearly used the appropriate methodologies for the different aspects of reality, confident that they would all come together when he was done. There was no dichotomy between his scientific and non-scientific work. They were both part of the same search. An example of the way he saw the connection lies in his consideration of the ether. The ancient Stoic tradition maintained that the marvelous organization of matter and the variety of living things so well constructed for the purposes of life could not be the result of the simple mechanics of atoms. They therefore postulated a continuous material medium permeating the entire world, vital and active, whose action gave form to the world and its parts, and was, in essence, a universally present deity. The alchemists preserved and handed down this idea through succeeding generations to Newton, who received it as the ether, through which forces could be transmitted through space. The Stoic tradition has its echo in the Christian doctrine of an omnipresent God, which takes the "argument by design" as one of the proofs of His existence. It was the concern with the vital agent and its relation to God that distinguished alchemy from chemistry, which could study only the grosser transformations of matter.

Like other natural philosophers of his time, Newton had a problem with action-at-a-distance. It was hard to admit that one body could act on another without anything between them. In his purely scientific work he refused to discuss the method by which gravity works ("I do not make hypotheses".), but he was concerned about it and thought that gravitational force, as well as light and heat, could be transmitted by the ether, which was involved with the vital agent in alchemy. But he had reservations, and his published work about the ether was always qualified. Most notably, the issue of the ether was posed in his *Opticks*, in which it was presented in the form of questions rather than facts. From experiments on pendulums he concluded that, if there were an ether, it did not retard their motion. Also, it would have self-contradictory properties, such as a density much less than that of air and yet an elasticity very much greater and therefore like an extremely dense solid. The contradiction later became extreme when light was found to consist of transverse waves and therefore had to be like a solid. Although Newton was ambivalent about the ether, he was fascinated by it as were many who followed him. How could anyone not be fascinated by a world in which one of two wildly weird things must be true? One: bodies act on each other by forces that act through a space that contains *nothing at all*; or two: space is filled with an invisible, undetectable material with self-contradictory properties unlike any that has ever been seen. Each possibility is beyond common sense and each poses an unfathomable mystery.

The connection of both natural philosophy and alchemical vitalism to Newton's theology is clear, since he saw it as the expression of the continual work of God's Providence in the world. Like the other seventeenth-century scientists, he saw no contradiction between religion and science and believed that the major purpose of science was to show the details of God's work in the physical world.

At a deeper level, Newton urgently pressed the "argument by design", using all he knew about physics and astronomy to declare that existence could not arise from nothing and could not be organized or maintained without the direct action of Divine Intelligence. And he used gravitational arguments to claim that the universe was infinite, in keeping with the infinitude of an omnipresent God. For, if the universe were finite, all the matter in it would long ago have clumped into a single mass at its center.

This was his mysticism: that the universe was truly awesome with some deep significance that could be ascertained by the power of the intellect. It was wrapped up with God and spirituality, and Newton rejected neither revelation nor meditation; and yet, without realizing it, he was the model of the modern scientific mystic, who is struck by the mysteries of the existing world, who believes that science is the only method of knowing that is not totally inadequate, and who wants not only to experience these mysteries, but to understand and be immersed in them.

His desire to understand included everything there was. He knew that his abilities were phenomenal, and yet to the modern mind it requires a special hubris to think that one man could find the inner secrets of *everything*. He was unique in three ways: he had a powerful intellect; he was tenacious and patient, never letting go of a problem until it was solved; and he acted on his strong belief in the total unity, taking everything in science, religion, and mysticism as part of his search. "He regarded the universe as a cryptogram set by the Almighty". (Keynes). Newton's confidence in his intellectual powers went beyond the point of arrogance, and he fully believed he could decipher the code. His attempt to do so was the hot, ruling passion of his life. His mission and his ego are best summed up by quoting White (page 331):[7] "Newton maintained an obsessive belief in his own uniqueness: he was convinced there could be only one Christ-like interpreter of divine knowledge in the world at any one time, and he never doubted that he was the chosen one". The intensity of his studies was driven by this passion and he did not care if others knew of his work, at least initially. He was secretive and published reluctantly, only at the urging of the few friends who knew something of his towering accomplishments. Anyone familiar with this part of his personality would conclude that he cared only for the knowledge and for nothing else. But his ego was as huge as his genius and when his work became known and brought forth the inevitable criticisms and questions of priority, he reacted strongly and became embroiled in bitter feuds. These had the effect of delaying progress in science and mathematics and were totally unnecessary because all those involved had accomplished so much that their fame was assured, regardless of any questions of priority.

Anger and resentment swirled around Newton's three greatest achievements: optics, planetary motion and the calculus.

His two greatest antagonists could not have been more different. While Robert Hooke was in his teens his father committed suicide, leaving his son neither money nor social position, and Hooke worked as a servant to get through Oxford. To support himself and have some opportunity for research, he took a job as assistant to Robert Boyle, the greatest chemist of the era, and later as Curator of Experiments for the newly formed Royal Society. The duties were

[7] See Additional reading.

not light and the quality and range of his work attested to Hooke's industry as well as his scientific brilliance. He had an insightful imagination, although he was an indifferent mathematician.

Hooke's early years were somewhat similar to Newton's, but his innate character was very different. He was gregarious, loved a good time, worked in many fields, nearly to the point of being dilettantish, and was prone to exaggerate his accomplishments.

G. W. Leibniz, on the other hand, was the son of a professor, a diplomat and philosopher, and a great mathematician, with the only intellect that could be said to approach Newton's. He was born in Leipzig in 1646 and spent most of his life in various kinds of diplomatic and governmental service. In 1676 he became the ducal librarian in Hanover, and this gave him the time to pursue his philosophical and mathematical interests.

Newton's problems started when, after being elected a Fellow of the Royal Society on the basis of his construction of a reflecting telescope, he submitted a letter describing his experiments with prisms in which he concluded that white light was a mixture of all the colors of the rainbow. As curator, it was Hooke's job to read all submissions and report on them to the Society. Newton's conclusions conflicted with Hooke's own, which were contained in his *Micrographie*, and he gave Newton's letter an unfavorable review. The ensuing scrap involved a number of people, made Newton and Hooke lifelong enemies, and brought out the worst in both men's characters. It extended to the theory of gravity and the planetary system, with Hooke making claims of priority regarding the inverse square law that he was unable to prove.

The famous conflict on priority for the development of the calculus had a longer-lasting and more profound effect. Initially, Leibniz accepted the idea that the two men had arrived at the calculus independently, and at first so did Newton. But events involving others played into Newton's messianic ego and created a schism between England and the Continent that lasted for many years and held back mathematical progress in England for over half a century because it would not adopt Leibniz' superior notation.

A third conflict, while not so momentous, was equally acerbic. John Flamsteed, the Astronomer Royal, was charged with making observations on the positions of the Moon, data required by Newton for his calculations on gravitation. Inevitable delays in getting the data, Newton's imperious attitude, and his failure to give Flamsteed proper credit fueled a long feud between them, which ended only when Flamsteed died.

The complexity of Newton's character was not limited to his attitude towards scientific recognition and fame. Perhaps surprisingly, he was efficiently practical and effective in worldly affairs. He was elected the Member of Parliament for the University of Cambridge to the Convention Parliament of 1689, and Cambridge representative to Parliament on the strength of his work on a mission to James II. The King wanted Cambridge to admit a Benedictine monk to a degree, without taking the normally required oaths, which a Catholic could not do. Newton and his colleagues successfully opposed him on behalf of Cambridge. He was reelected in 1701 when he was in London and it was politic for him to be in Parliament.

His life took a major turn in 1696 when he was appointed Warden, and later Master, of the Mint. He had suffered a nervous breakdown lasting for a good part of 1693 and it was feared that he had completely lost his mind. Newton had been one of a distinguished group of citizens that advised the King on what to do about the current monetary crisis, so he had some familiarity with the Mint and its problems. His friends helped him secure the position, in which his lifestyle was quite different from the solitary one at Cambridge, which confined him to his room and his laboratory. The Mint was in trouble and needed someone with technical expertise and a strong will to reconstruct it. Newton did not disappoint. He took up residence in a fine house in London and was soon immersed in its social and political life while vigorously attacking the problems at the Mint.

He displayed administrative abilities and political skills that were surprising in a natural philosopher who had spent most of his life in seclusion. Plans for a desperately needed recoinage had been thwarted by a combination of old, inadequate production facilities, incompetence, and corruption. The situation was out of control.

Newton revamped production facilities, got rid of the incompetent and corrupt, and successfully completed the recoinage. A search of old records showed him that he had powers that had not been used for many years, and, never being one to avoid the use of his authority, he zealously pursued, prosecuted, and jailed counterfeiters. To this end, he created a large network of informers, whom he worked with and met in the seamiest, most dangerous parts of London, and he had himself appointed Justice of the Peace in seven counties so he could secure convictions and dispense sentences. He was personally outraged by counterfeiters and prosecuted some of them right to the gallows.

His work at the Mint is credited with literally saving England from financial ruin and public riots. The coinage had become so debased by clipping and counterfeiting that people could no longer carry on business and were being driven to the barter system used in the darker years of the Middle Ages.

Who would not have been content with Newton's accomplishments and status? Recognized as the greatest intellect who ever lived, wealthy from wisely investing a modest inheritance and his salaries from lucrative positions, known as the savior of the English economy, knighted in 1705, he had every reason to enjoy his later years, become more mellow, and bask in the light of public adulation. But his personality dictated that it was not to be. In his own family he played the part of generous patriarch, helping little-known relatives, some of whom became visible only because of his fame. And he was good enough to those who admired him and paid him the homage he felt was his due. But he remained jealous of anyone who did not recognize his right to power and privilege. He was scheming and vindictive to the end, especially in his role as President of the Royal Society, which he ruled with an iron hand and where he rewarded his admirers and punished those who disagreed with him.

Newton was a most complex man: mathematical physicist, alchemist, theologian, Biblical scholar, administrator, and public servant, a loner and an astute seeker of power and status.

Many people have thought about gravity, motion, and the heavenly bodies, and Newton was neither the first nor the last scientist to wonder at them. But he was the one who first made sense of it, put it in terms that could be tested, and provided the platform for future studies and a deeper understanding. Only Newton could have successfully solved the problems of planetary orbits by creating the entire theory of mechanics needed to do so, and also to apply it to a huge range of celestial and terrestrial phenomena.

2

The giants

Of all the awe-inspiring sights of nature, nothing compares with the star-filled heavens on a clear night. Haze and artificial lights now obscure this for many people, and one must go to relatively pristine locations, such as the Arizona desert, to see the majestic, overwhelming vision that is the night sky. That black dome, studded with sparkling lights, with the anomalous and variable Moon, is a commanding presence with an almost tactile impact. It defines the universe, an enormous, wondrous thing that cannot be denied, reaches into our very soul, and demands explanation. So it has been ever since people looked up, and so it is today.

Looking out over Athens from the Acropolis, in the shadow of the unparalleled sublimity of the Parthenon, to the surrounding mountains, makes it easy to believe that the higher human consciousness arose here. It is the center of the origin of Western thought: in philosophy, literature, drama, art, and, above all, in mathematics and science. Certainly they had predecessors, but the ancient Greeks were the first to apply reason and careful observation thoroughly, continually, and systematically in attempts to understand the composition and workings of the cosmos. They explored a number of theories, including a heliocentric system having many elements in common with modern knowledge. These were not just imaginative speculations; they were subjected to careful analysis in accord with observation and rigorous logic. The theory that survived, and was later integrated with Christian theology, was the Earth-centered system of Ptolemy, consistent with the physics of Aristotle. It was based on experience and logic, and was successful in that it was a coherent picture of the entire universe and enabled computations of the positions of the Moon, the Sun, and the planets that agreed with astronomical measurements.

Aristotle, Archimedes, and Ptolemy are rightly regarded as Hellenic giants. Two of them have been labeled as obstacles to modern science because their followers became little more than petrified parrots, parsing each sentence of the ancient texts to fit into a preconceived religious framework. They and their contemporaries would have been surprised. Both of them were innovators of the highest order and together created the traditions of science and astronomy that led to Copernicus, Brahe, Kepler, Galileo, Newton, and Einstein.

Aristotle took all of knowledge as his turf and spent his life studying and lecturing on subjects ranging from ethics and logic to botany and cosmology. He was well born. His father was physician to the King of Macedonia, and, although the tradition that he was Alexander the Great's tutor is doubtful, he certainly spent a lot of time at the Macedonian court and knew Alexander's father well. He was an academician all his life, starting with his entrance to Plato's Academy in Athens in 367 BC at the age of seventeen and culminating in the founding of his own school, the Lyceum, in 335 BC.

Perhaps Aristotle's most important contribution to science was his rejection of the Platonic "Ideals". Plato saw that everything was always changing and could not believe that such a confusing flux could represent the underlying reality of the world. Instead he held that the true reality is a set of idealized concepts on a higher and more exalted plane than what we observe.[8] The "Ideals" arose from geometry. There were physical objects that, for example, looked like squares, and any number of squares could be constructed

[8] The idea that there is a reality hidden by what we observe is far from dead. Ahab's obsession was driven by his belief that the world of the senses is just a veil masking a malevolent reality. Hunting the white whale that took off his leg was his way of striking back at that hidden reality. He failed.

from available materials. But none of these were perfect. On close inspection, their angles were off, or one of the sides was not exactly a straight line, and even when measurement showed that the object was quite regular, everyone knew that there would always be some deviation from perfection. So no physical square was a *true* square. A true square was an ideal geometric concept that could never be realized in practice. So Plato held that only the ideal concept of the square, as defined in geometry, had "real" existence. This is easy to accept as the definition of geometric reality. But this idea was expanded to include the rest of the world. Thus, we can see a large number of horses, all of them different and all with some imperfection or other, but the reality is the perfect idea of a horse, while the observed horses are imperfect manifestations of the idea of "horseness".[9] Aristotle, however, took change as a reality and held that there is an explanation for every change, thereby making it reasonable to look into motion and its causes.

Ptolemy (AD 85–165) came along about 400 years after Aristotle. He was an eminent mathematician, found a formula that gave π accurately to three decimal places, and proved the theorems he needed for his astronomical calculations. These amounted to significant advances in trigonometry. He made observations of the heavens at Alexandria for nearly fifteen years, which he used, along with data from predecessors, to calculate the orbits of the planets and the Sun. Of course, he assumed that the cosmos was geocentric because that was in accord with observation and with the natural philosophy of Aristotle. The concept of geocentricity was not original with him. The Earth was always thought to be the center of everything, and in fact the notion of an Earth-centered world had been developed by

[9] I must confess that I do not really understand what I just wrote, since the independent existence of an idealized concept is alien to me. I have read some of what has been written about Platonic ideals and have concluded that I live with such a different world view that I cannot communicate on this subject. My problem is that Plato seemed to believe that his "ideals" had an objective reality.

Plato and his pupils. But other theories existed. About a century before Plato, Pythagoras adopted a cosmological model that was not geocentric. He abandoned the ideas of the Earth's immobility and centrality, and held that the Earth and the heavenly bodies all rotated around some point (the central fire), which was neither the Earth nor the Sun. This was abandoned because it was not supported by observation, whereas all observational evidence pointed to a stationary, centrally located Earth.

Ptolemy's calculations described a detailed system of the universe, which he published in the *Almagest*. (The title comes from an Arabic word meaning the "Greatest Compilation", which is a promotion of Ptolemy's original, and more modest, title, "The Mathematical Compilation",. Like much other mathematics and science, the ancient Greek knowledge entered Europe via Latin translations of Arabic scholars, who acquired important parts of their knowledge from India.) The *Almagest* was beautifully organized and well written, so its contents were readily available to people with a classical education. It was a considerable advance over Aristotle's idea that the Moon, the planets, and the stars all moved in perfect circles, and was the standard astronomy text for a millennium and a half. Ptolemy also wrote important texts on geography, which were an attempt to map the known world, and on optics, in which he reports his studies on color, reflection, refraction, and mirrors. His optics is a work in which mathematics and experiment are combined in a manner worthy of any modern scientist.

The Greek accomplishment cannot be overestimated. Starting from practically nothing they developed the methods of logic, mathematics, observation, and experiment that laid the foundations for every science. This legacy could have grown into a rapidly expanding store of knowledge that would have made European history very different by advancing civilization much faster. However, the gradual decay and ultimate fall of the Roman Empire delayed progress for a millennium. Western Europe was cut off from the old Greek sources, and the study of nature focused on its relation to religion, not on natural phenomena themselves.

There was practically no access to the older achievements, and scholarly work was restricted to the efforts in some monasteries to preserve what could be salvaged of the ancient knowledge.

It was not until the tenth and eleventh centuries that the Greek texts started to get to Western Europe. They were preserved and studied by Arab scholars, who also absorbed the great Indian advances in mathematics. The Arabs transmitted some of the most important mathematical concepts, such as zero, a notation that defined integer values by position, summation of series, and algebra to Europe, as well as much of the science, philosophy, and literature of ancient Greece.

In view of the great conflict between Galileo and the Catholic Church in the seventeenth century, it is interesting to note that the Church did not readily adopt the Ptolemaic system. In fact, it was not until the great unification of Catholic doctrine and Aristotle by St. Thomas Aquinas in the thirteenth century that it was fully accepted. The objections to the Ptolemaic cosmology arose from the desire of the early Church to purge itself of all taints of paganism. Greek knowledge was therefore ignored and actively opposed until its great synthesis with Catholic dogma by St. Thomas Aquinas.

The true intellectual Greek ancestor of modern science is Archimedes (287–212 BC). He was certainly one of the greatest mathematicians of all time. He was military advisor to King Heiro, and his use of catapults, magnifying mirrors, and huge claws to sink, burn, and capsize ships in defense of Syracuse is legendary. Archimedes' machines were so effective that many Romans were convinced they were facing a great magician with supernatural powers. At last, using overwhelming force, the Romans took the city, and Archimedes' was killed by a soldier during the looting that followed. Two memories of Archimedes have persisted to the present day. The first is his discovery of the law of buoyancy, known to all students of physics as Archimedes' principle, and the second is his excitement at finding the solution to the King's suspicion that a metal worker had cheated him from some gold that was to go into a new crown. The flash of insight experienced by someone trying to solve a tough problem is now called the eureka moment, in recognition of Archimedes' shout when the answer to the question of the King's crown occurred to him. This is the first recorded instance of that thrilling excitement at discovering something remarkable and important about the real world. The story of Archimedes jumping naked from his bath and running through the streets shouting "I've found it"! resonates with every scientist who has been fortunate enough to learn something new.

Archimedes' legacy is quite different than that of Aristotle, Plato, or even Ptolemy. He addressed specific problems of mechanics and mathematics, was not averse to experiment or careful observation, and did not try to construct grand, overarching systems of thought. His contributions to the theory of mechanical statics, hydrostatics, and the simple machines laid the foundations for future classical mechanics, and his mathematics was far ahead of its time. He was the first to define and use a geometric series, found more accurate calculations of π, was a superb geometer, and anticipated the integral calculus by calculating areas bounded by lines and curves using methods of successive approximations.

The Aristotelian–Ptolemaic tradition fit well into Christian doctrine because it made Earth, with its birth of Christ, the center of all, and agreed with cosmological references in the Bible. In spite of being wrong, it was tenacious, and the correct heliocentric theory of planetary motion against a background of fixed stars was not accepted until the seventeenth century.[10] An important element of this tradition was that Aristotelian physics was not mathematical and was concerned with finding qualitative causes, while Ptolemaic astronomy, which ignored causality, was highly mathematical and its major purpose was calculating stellar

[10] Of course, from a rigorous viewpoint, both the Ptolemaic and the heliocentric reference systems are equally valid. I call the heliocentric system "correct" in the sense that it is simpler and easily leads to new results and more accurate descriptions. An important example is that the Ptolemaic system claims that the Earth is the absolute center of the universe.

and planetary positions. The dichotomy was a serious one and a matter of basic principle rather than a simple result of the fact that Aristotle did not do mathematics. He did not think that mathematics had anything to say about nature. He thought that the purpose of physics was to find *why* nature behaved as it did, and this had nothing to do with mathematics. Astronomy, on the other hand, was concerned with finding the positions of celestial objects, primarily for constructing a calendar and casting horoscopes. So mathematics was required as a practical matter but had nothing to do with causes. Causes were unnecessary in mathematical astronomy because everyone knew that the fundamental motion of celestial objects had to be based on circles. Astronomical systems were thereby merely computational aids, not having much to do with trying to understand the essence of reality. Aristotle reconciled astronomy and physics by ascribing a teleological character to both terrestrial and celestial objects. This was an essential part of his theory of matter, which maintained that there were four basic elements—earth, water, air, and fire—and that all matter was some combination of these.[11] Each element had a different tendency, which apparently was controlled by its density.

The concept of gravity as a force was not a part of either the physics or the astronomy of the Greeks. Of course, they wondered about the fact that objects fell to the ground, and they had a simple explanation. Everything had a natural place in the universe that it tried to get to, and the natural place for dense objects was down. Fire, on the other hand, rose up because it had a divine character and its natural place was in the heavens. There was an equally clear explanation for Ptolemaic astronomy. The most perfect kind of motion was circular, and celestial bodies, being much more perfect than corrupt terrestrial matter, sought to move in circles.

[11] The four-element theory was important as a precursor to the acceptance of atomism. It was most probably not of Greek origin since it appears in several religious traditions of India that predate Aristotle. It could have been brought to the West by Pythagoras, who was said to have visited India.

When actual observation showed that planets did *not* move in circles, the Greeks described the motion as a compound of little circles going around a point on larger circles. These were epicycles on the cycles.

Things acted as they did because they had natural tendencies and were after particular end results. It was a teleological view, compatible with a Christian theology in which God's purpose is all-important. It was also compatible with the idea that there was a better place and that this place was "up", beyond the limitations, corruptions, and decays of daily life on Earth. Where else could such a place be than beyond the visible heavens, which were pure, perfect, and incorruptible compared to Earth, and where else could hell be, but below us, down, down, down, where evil and corruption found its ultimate expression? Science was so entwined with theology and ancient beliefs that natural truths could not be unraveled for centuries.

This changed within a span of about one hundred and fifty years, straddling the sixteenth and seventeenth centuries, when a small group of men created modern science and transformed the world. It was the beginning of a new kind of mysticism, deeper and more wondrous than anything previously conceived, based on the desire to learn and get behind the secrets of nature and to participate directly in the intoxicating mystic experience through true knowledge.

It was a heady time, full of intellectual ferment and liberation. It stood between the Renaissance, which prepared the way for the ascendancy of rational science, and the Enlightenment, with the succeeding centuries of ever-increasing progress in understanding the natural world.

It was still a totally religious time in which the churches wielded enormous authority and influence. Demonology, astrology, witchcraft, magic, and alchemy were still an integral part of much intellectual thought, even among some of the most enlightened. But mathematics, experiment, and logic were increasingly being recognized as critical for scientific understanding, and significant progress had been made in natural philosophy, especially in astronomy and mechanics. Aristotle's physics was already being rejected and Ptolemy's astronomy was dying. The men responsible for the greatest scientific advances were deeply religious and thought of their work as glorifying God. Yet they created the framework that weakened the power of organized religion, which, along with the conservative Schoolmen,[12] was fighting a rearguard battle against the new science. The age was ripe for Newton, and he came along at exactly the right time.

Newton had a lifelong feud with Robert Hooke and it was in a letter to Hookes, describing an argument about optics, that Newton made his famous statement that if he had seen further than other men it was because he had stood on the shoulders of giants. (Some historians point out that this could have had a double meaning because Hooke was a short, almost dwarfish man, and Newton was often acerbic and even insulting to his enemies.) Chief among the giants were Copernicus, Tycho Brahe, Johannes Kepler, Galileo, and Descartes. They truly were giants, without whom Newton could not have accomplished his astonishing revolution.

Newton was in possession of analytic geometry, a detailed heliocentric description of planetary motion, the laws of falling bodies, the idea of inertia, and the beginnings of the laws of motion, including the insight that acceleration is produced by a force. These were the products of the work of many mathematicians, astronomers, and natural philosophers, some of whom stand out because their contributions were so important, and in a direct line to modern views. They were passionate men,

[12] The term "Schoolmen" is often used to denote those academics, particularly those in non-scientific fields, who stubbornly upheld the Aristotelian–Ptolemaic–Christian corpus even while strong evidence was accumulating for the new astronomy and new mechanics. Note that many of the great innovators of the sixteenth and seventeenth centuries were academics with professorial appointments at the best universities. These included Kepler, Christopher Wren, Galileo, and Newton. Thus, while the universities were strong conservative forces, they harbored the initiation and growth of the new sciences.

imbued with the new scientific spirit and possessed by a desire to dig into the mystery of nature and expose its workings.

The Ptolemaic system was ingenious, complete, and in accord with the Aristotelian physics of the time. In fact it was an improved and quantitative version of the Platonic–Aristotelian cosmology, with added details to account for the observed motions of the planets.[13] The Moon, the five known planets, and the stars were thought to be embedded in a succession of "crystalline" shells rotating around the stationary Earth, thereby giving the heavenly bodies their orbits. By crystalline, they meant a solid glasslike, transparent material because the shells had to hold the heavenly bodies but were not visible. It was a geocentric theory because observation clearly showed that the Earth was stationary and the heavens revolved around it. The inner shells derived their motion from the outer ones. The outermost sphere, holding the stars, moved all the others and was called the prime mover beyond which was *nothing*. It was a closed universe and it was a small universe because the successive shells were close together. This system was adopted by the Church and extended to place heaven beyond the last sphere.

The first successful attempt to displace Ptolemy actually antedated the observational results that demonstrated the rotation of the Earth on its axis and its revolution around the Sun. Copernicus was a distinguished physician, a translator of ancient Greek writings, a public servant, an ecclesiastical administrator, and an expert on finance, writing an important book on monetary reform, which was highly acclaimed in his native Poland. Astronomy was his passion, and although it was a part-time avocation, he managed to continually perform astronomical observations from his garden. He was far from a revolutionary

[13] The word "planet" was originally applied to all heavenly bodies that moved and included the Sun and the Moon. They were "wanderers" as opposed to the fixed stars that never altered their positions in the sky. I am using the word in the modern sense, in which the Sun and the Moon are not called planets.

and arrived at his conclusions from the accepted humanistic practice of looking into ancient Greek texts to feed the expanding body of knowledge. He stated that he was led to his system because different astronomers gave different calculated results, so he searched the Greek writings for alternatives to Ptolemy. The complexity of the final version of the Ptolemaic system surely contributed to these discrepancies. Perfect motion was circular motion, and to preserve circularity while achieving agreement with known observations, about forty cycles and epicycles were progressively added to the original simple, circular orbits. Copernicus thought this was too cumbersome and overly complicated. He found another possibility in the Aristarchian tradition of a heliocentric system, which had lost out to Aristotle because a moving Earth was so obviously ridiculous. The heliocentric theory was simpler and more elegant to Copernicus, whose major work, *De Revolutionibus Orbium Coelestium Libri Sex* (*Six Books on the Revolution of Celestial Orbs*) was published in 1543, the year of his death. It closely paralleled the structure of Ptolemy's *Almagest*, carefully refuting Ptolemy's individual claims for its validity and showing that the heliocentric system could address all the known astronomical issues. Furthermore, it gave all numerical results with greater accuracy and simplicity than was possible in the Ptolemaic system.[14]

Copernicus' position in the history of science illustrates an important point. Every major advance has precursors and has been anticipated to a greater or lesser degree by past and contemporary colleagues. But the one whose name gets attached to that advance is the one who does something special. It was the great care and thoroughness of *De Revolutionibus* that showed how Copernicus was in a different league than others who had thought about a heliocentric system. he did not simply state an opinion; he gave

[14] It is true that Copernicus had to retain some of the Ptolemaic devices, such as epicycles, to achieve accuracy, but to a much lesser degree than Ptolemy. But to Copernicus, the main reason for adopting the heliocentric system was its rationality and simplicity.

cogent reasons based on logic and simplicity, he gave a careful scientific critique of the Ptolemaic model, and, most importantly, he demonstrated that his system gave practical computational results of greater accuracy. He did this in at least as great detail as the classic presentation in the *Almagest*. His masterful scholarship, his cogent logic, and the success of his calculations could not be ignored. He was singular. Also, his results initiated an exponentially growing body of work by others who accepted, developed, and disseminated the heliocentric model. Similar conditions identified the great achievements in astronomy and mechanics associated with the names of Galileo, Newton, and Einstein.

The Copernican system had implications far beyond the acceptance of the Earth's motion and the lack of any single center of the universe. One was that there was no distinction between heavenly and terrestrial matter. Aristotle held that the material below the crystalline sphere of the Moon was different from the heavenly, more perfect, stuff constituting all that was beyond this sphere. Such a distinction was a barrier to the idea that any knowledge about moving bodies on Earth could be applied to celestial motions. Another implication that Copernicus had to accept was that the fixed stars were very far away. If the Earth moved around the Sun, then the stars must exhibit a parallax shift when they are observed at different times of the year. Parallax is just the change in apparent position when an object is viewed from two different places and is easily demonstrated by just walking a few blocks while looking at a tree. The tree looks as if it is in different places as you walk. In the beginning, it may seem to be in front of the Moon, but a block later, it seems to be to the left. That is parallax. In the same way, the fixed stars when observed in the winter would be displaced from the positions seen in the summer if the Earth had changed its position during the year.

The tree looks like it changes position only if you are walking. Similarly, there would be no parallax for a stationary Earth, and the fact that none was observed was taken as evidence for a geostatic model. But the parallax shift is smaller for objects that are further away. During our walk at night, we see that the apparent position of the Moon against the fixed stars in the sky remains unchanged. The apparent shift in position of the tree and the Moon as we walk is so different because the Moon is much further away than the tree. Similarly, if the Copernican system were true, the parallax shift had to be too small to be detected, and this meant that the distance from the Earth to the stars had to be very great, much greater than common wisdom would accept. The heliocentric picture was the first step in recognizing the immensity of the universe.

The idea that the laws of nature were the same for celestial bodies as for matter on Earth was essential for the progress of science and for an understanding of planetary motion. In the heliocentric theory, the Earth was just another planet going around the Sun, so it was hard to keep the old idea that terrestrial matter was different from that in the heavens. And it became easy to believe that the same dynamical laws hold for planets as for bodies on Earth.

Also, Copernicus' heliocentric model implicitly brought out the issue of gravity. In Aristotelian physics and Ptolemaic astronomy, objects on Earth fell because they sought the center of the universe, which was also the center of the Earth. But if the Earth was no longer the center, why should objects fall? This question did not need to be asked before Copernicus.

Copernicus erased the barrier between physical causes and mathematical representation that had been a staple of Aristotelian physics, and he presented a choice of mathematical theories based on observation and reason. His work was truly revolutionary but was not immediately recognized or adopted as anything more than a mathematical scheme. The Church, which still accepted the Aristotelian separation between mathematics and physics, did not object to a heliocentric system, provided it was described as a hypothetical model useful for calculations.

It was Galileo's insistence that a heliocentric model was literally true that got him into trouble with the Church. If he had agreed to present it as a hypothesis or a mere method of calculation, he would have had no problem.

Copernicus waited thirty years before finally publishing his complete work in 1543, the year of his death, but his theory was known in the astronomy community long before then. He had distributed a handwritten manuscript, called the *Little Commentary*, which laid out the axioms of the heliocentric system, and in 1540 his disciple Georg Rheticus published a book called *The First Narration*, which was an introduction to Copernicus' work.

He surely believed his model to represent literal truth, and he knew that this was in direct conflict with the teachings of the Church, which was committed to an Earth-centered astronomy. But his delay seems to have been as much a concern about ridicule and criticism as a conflict with religion since, for most people, the idea of a moving Earth was absurd. In fact, the Church at that time welcomed new advances in astronomical calculations because it urgently needed more accurate ways of constructing calendars so as to properly date religious holidays. These were to be regarded merely as mathematical methods that had nothing to do with reality.

To avoid conflict, an anonymous preface, replacing that of Copernicus (added by Andreas Osiander because of the well-known opposition to the Sun-centered theory) stated that the theory put forward in *De Revolutionibus* was only a mathematical hypothesis and not to be taken literally. In fact, that was how the Catholic Church first regarded the book, and it was not condemned until seventy-three years later in 1616 in connection with the contentious Galileo affair. Four years later the book was "corrected" after some sentences that presented the heliocentric theory as fact, rather than a mere mathematical hypothesis, were removed or changed. Yet it stayed on the Index of Forbidden Books until 1835.

Both the diurnal rotation of the Earth on its axis and an annual revolution around the Sun were hard to accept, and, as Copernicus' ideas spread, a great debate took place between the Copernicans and the Ptolemaists that was not settled until the work of Kepler and Galileo. Even then the idea of a moving Earth was vigorously opposed. The opposition was not merely that of hidebound conservatism. Within the scientific and astronomical context of the time, there were sound reasons for objecting to a moving Earth. First, there was Aristotelian physics, which had been very successful in organizing scientific knowledge. Dense objects had a natural, innate tendency to move towards the center of the universe. The Earth was a dense object so it was at the center. Furthermore, it was extremely unlikely that an object as dense and heavy as the Earth could ever move. Motion of the Earth in a circular orbit would violate the natural order, which required celestial objects to move in circles and terrestrial objects to move in straight lines. These were objections based on the existing scientific paradigm, but there were also objections based on observation and experiment. If the Earth rotated, the atmosphere and all objects not tied down would be stripped off into space, just as bits of clay fly off a potter's wheel. Also, a stone dropped from a tower would not fall at the tower's base; because of the Earth's rotation, it would fall some distance away.

Copernicus anticipated these objections and met them by pointing out that: since the Earth is a sphere, it could also have a circular orbit similar to the celestial spheres; also, it was more likely that the Earth would rotate than that the entire massive celestial system would revolve around the Earth, which would require truly incredible speeds; and the atmosphere and all objects on the Earth are part of the Earth and rotate with it, so there is no tendency to fly off.

History was on the side of Copernicus, but he provided no clue to the existence of gravity. In the contemporary paradigm, objects fell to Earth because it was in their nature to do so and the celestial bodies stayed in their places because they were embedded in the hard, transparent celestial spheres.[15] Copernicus included these spheres in his model, so he had no occasion to

[15] One may wonder why the Aristotelians did not ask what held the celestial spheres in place. They did not because they assumed that everything had a natural place and a natural motion which they possessed as a condition of their existence.

consider any force required to hold the planets in their orbits. Nevertheless, removing Earth from a privileged position as the center of the universe was an important step for astronomy and physics. Terrestrial and celestial matter could no longer be divided into two distinct kinds. Equating celestial and terrestrial motions destroyed the Aristotelian reason for objects to fall; some other reason was required.

A most revolutionary result was that the Earth no longer provided an absolute reference system. In fact, Copernicus discussed relative motion, using a ship leaving port as an example. Galileo later used the same example to discuss relative motion and the concept of inertia.

His reluctance to publish was in accord with the Pythagorean tradition in which advanced knowledge was thought to invite ignorant travesties among those who were not specially trained. The danger was real because all educated people of the time felt competent to understand and judge any science, particularly astronomy, even if they had no specialized background.

The fact that such a purportedly conservative, tradition-bound mind was able to accept such a revolutionary idea shows that Copernicus was more modern and more daring than is commonly supposed. He still retained the concept of the circle being the most perfect curve, so he had to introduce some epicycles, many fewer but similar to those in the Ptolemaic model, to reproduce observed planetary positions. Also, he assumed that the center around which the planets revolved was not precisely at the Sun but a little displaced from it, thereby anticipating to some degree the correct elliptical orbits.

But Copernicus was right about how the book would be received. While using his system for purposes of calculation, the idea of a moving Earth was too preposterous to be taken as a fact, and was rejected by most. At first, only rabid anti-Aristotelians accepted it as a literal truth and used it as a weapon against the old order.

An important aspect of Copernicus' work was the combining of mathematics with physics. The Ptolemaic system was thought of as a mathematical aid to calculation, not integrated with Aristotelian cosmology, which was purely qualitative. No one had connected astronomical mathematics with physical reality as closely as Copernicus. And yet there was no compelling empirical proof that his system actually did represent physical reality. That had to await the unlikely team of Tycho Brahe and Johannes Kepler.

Giordano Bruno is often depicted as a player in this game because he was an avowed Copernican and was burned at the stake in 1600. He certainly spoke for the Copernican system with passion and conviction at every opportunity and was effective in publicizing it. Primarily, he used it as a justification for a form of Pantheism, thereby proving that those who worried about the misuse of good science were correct. However, Bruno was a thoroughgoing rebel on almost every front and was reluctantly burned for a variety of religious heresies, only one of which was his astronomical beliefs.

The era of modern empirical proof for the heliocentric system started with Tycho Brahe, who was in a unique position to advance astronomy. He was a nobleman, born into a wealthy, influential family with close ties to King Frederick II of Denmark and marked from birth to take his place in public life among the high aristocracy. But he had other plans. He was committed to astronomy in his youth and made his first observations, with improvised instruments, in his teens, and by the age of sixteen he was keeping a log of his work. His measurements on stellar positions and on the conjunction of Saturn and Jupiter were in disagreement with both existing sets of astronomical tables, one based on Ptolemy and the other on Copernicus, thereby convincing him that more accurate data were essential. Tycho was drawn to the stars by astrology and was also absorbed in alchemy, but the appearance of a new star in 1572 (a supernova resulting from the explosion of a white dwarf star) drew most of his energies to astronomy, although he maintained an interest in alchemy all his life. In 1573 he published *De Stella Nova*, which described his observations and conclusions of this unexpected phenomenon, which had no place in the Ptolemaic system. The book greatly advanced his reputation and also declared his vision of himself and his destiny, rejecting and belittling the life of the nobility and pledging himself to the pursuit of knowledge. The new star was a sensation because it was the first direct evidence that the heavens are not eternally unchangeable. Many observers insisted that the nova was a sublunar phenomenon, and since terrestrial matter included everything beneath the lunar sphere, the separation between Heaven and Earth was maintained. But Tycho new better because his measurements were accurate and he trusted them. It was a crack in the Aristotelian–Ptolemaic perfection. The crack widened when Brahe studied comets and concluded that they had trajectories that went far beyond the Moon and crossed the orbits of planets. This destroyed any possibility that crystalline spheres existed (a belief never held by Brahe). Now the question of what held the planets in orbit had to be confronted.

Tycho Brahe was a true rebel who was ardent and self-confident to the point of arrogance. When he was barely out of his teens, he got into an argument with another wild youngster that ended in a duel in which he nearly lost his life. A saber slash across his face left a long scar and took away a part of his nose, a disfigurement Tycho tried to correct by making a part from an alloy of silver and gold, mixed to approximate the color of flesh. He attached this to the remains of his nose with an adhesive ointment.

His independent rebelliousness was evident in his marriage to a commoner and this caused him some trouble because she was not permitted to the many functions, such as receptions, parties, weddings, and funerals that were an important part of the aristocratic scene. Tycho attended these alone.

His close relationship to the King gave Tycho the most remarkable observatory ever built. It was on the cliffs of the island of Hven, just off the Danish shore, a gift from Frederick II, along with enough money to build a fairy-tale palace, turreted and domed. It had movable roof sections for observing the sky, rooms for many assistants, a large library, instrument shops, and alchemical laboratories. It was a vision straight out of Disneyland.

He named it Uraniborg, after the Muse of astronomy, and it quickly became the pre-eminent observatory of Europe.

Measurements of stellar and planetary positions are useful in themselves, and these were the most important parts of Tycho's work because they led Kepler to his three laws of planetary motion, which Newton used as basic data for his work on universal gravitation. At the time, Tycho's observations were of major interest to others in addition to astronomers, especially astrologers. Astrology was a flourishing enterprise, favored by kings who had to make many decisions on the basis of uncertain information, and accurate planetary positions were essential for accurate predictions. Tycho's own attitude was that the stars certainly did influence human affairs, but the predictions were not absolutely deterministic; they could be modified and even completely changed by the exercise of free will.

Tycho's astronomical work was prodigious and not restricted to just recording observations. The observations were of unprecedented accuracy because he developed new instruments of such precision that the errors in previous measurements were reduced by a factor of fifty or more. Previous measurements had errors in the range of minutes of arc; Tycho reduced these to seconds of arc.[16] He described these instruments, their use, and construction, in a widely admired book.

He carefully studied the comet of 1577 and concluded that it was orbiting the Sun and was far beyond the Moon. This confirmed his belief that the Ptolemaic model was wrong and reinforced his ambition to become the greatest astronomer in history by finding the correct model that would replace Ptolemy. He did construct such a model, but in spite of his iconoclastic, venturesome nature, he could not bring himself to abandon a motionless Earth and accept Copernicanism. The Earth could not move; all the evidence was against it. The Tychonic model, had the five planets circling

[16] Measurements of the angular separation of celestial objects were the only quantitative data that could be made at the time. Spectral analysis was not available until much later.

the Sun, but the Sun and the Moon circled a stationary Earth. The Tychonic system had several advantages. It was just as accurate as that of Copernicus for astronomical calculations, accounted for all known observations, and did not have that controversial and troublesome feature of the Copernican model, a moving Earth. Tycho therefore retained what was thought to be the best of the traditional view, while providing a straightforward and accurate calculational procedure. Because of this, the Tychonic model gained a wide following and was a serious competitor to that of Copernicus.

Tycho's multiple and ambitious projects were usually pursued with the help of assistants, and he was always looking for young talent. He had been impressed by some of Johannes Kepler's early ideas and in 1600 admitted him to his observatory in Prague, which Tycho occupied since leaving Denmark in 1599. At the age of 54, Tycho Brahe was the most distinguished astronomer of his time, while Kepler, at 29, was in the early stages of his career. It took some time before the terms of Kepler's appointment were concluded, but he was on the road to his three laws of planetary motion and immortality.

Kepler and Tycho were totally unlike both in background and in temperament. While Tycho was an aristocrat and moved in the highest circles with an extensive and often supportive family, Kepler was born into poverty with uncaring grandparents and a vicious brute for a father who hired himself out as a mercenary soldier and left home when Kepler was five years old. His mother was a nasty, ill-tempered woman who was actually tried for witchcraft and escaped conviction only through Kepler's intervention. Although he sometimes exhibited a temper, Kepler was modest and self-critical, while Tycho was egotistical and overbearing. Both men were religious, but Kepler was more so, and the object of his studies was to look into the Mind of God, while Tycho's major ambition was to become immortalized as the greatest of astronomers. Kepler was more of a mystic, as evidenced by the titles of two important books, *Mystery of the Cosmos* and *The Harmony of the World*.

Tycho Brahe was jealous of his data. They were the result of years of intense labor, and held the secret to knowledge, which Tycho believed he had found and embodied in his model of the solar system. He knew that Kepler had analyzed earlier data and concluded that the Earth went around the Sun, as in the Copernican model, in direct contradiction to Tycho's conclusion. So Kepler was deeply disappointed because, contrary to his expectations, he had very limited access to Tycho's data until after his death in 1601.

The precious numbers gave the positions and velocities of the planets throughout their entire periods, so detailed calculations of their orbits could be made. These led Kepler to assert that three major regularities existed for planetary motions that are embodied in three laws. These laws state that:

1. The planets move in elliptical orbits with the Sun at one focus. (1605)
2. As they move around the Sun, the planets sweep out equal areas in equal times. (1602)
3. The squares of the orbital periods of the planets are proportional to the cubes of the average distances from the Sun. (1618)

That Kepler could extract these laws from the volumes of Tycho's data is a testament to his industry and to his belief in a harmonious world. They are short and simple and were a turning point in astronomy that prepared the way for Newton.

The first law states that planets do not move in circles. They all have elliptical orbits with the same basic geometry, and this indicates that some fundamental common factor is at work.

Another deviation from the idea of a circular orbit was the fact that a planet moved faster when it was closer to the Sun than when it was further away. This is embodied in the second law.

The first two laws refer to individual planetary orbits, while the third law applies to the entire set of orbits, thereby introducing a regularity in the solar system as a whole.

Kepler arrived at his conclusions from a laborious and detailed analysis of Tycho's data, and Newton used them to get the quantitative expression of the law of universal gravitation.

The discovery of the laws of planetary motion was Kepler's crowning achievement, but his other accomplishments were far from trivial and had a wider range than those of Brahe. He made major advances in optics, as shown in his books *Astronomia Pars Optica* and *Dioptrice*. He worked out the theory of geometric optics based on light rays, including the concepts of real, virtual, and inverted images, explained how a telescope and a pinhole camera work, and correctly described image formation on the retina of the eye, as well as binocular depth perception, and showed that a parabolic mirror focused light rays to a point. He even designed spectacles. His mathematics included a study of logarithms and calculations of the volumes of solids, which anticipated the calculus.

Kepler came close to Newton's theory of gravitation. He assumed that all bodies attract each other with a force that diminishes with distance and, in a short story, which is probably the first science fiction, described correctly the point between the Earth and the Moon at which their gravitational fields exactly cancel. He had concluded that the intensity of light from a point source (such as a candle) decreased inversely as the square of the distance because the area of any sphere around the source increased directly as the square of the distance and the total illumination from the candle could be preserved only by an inverse square law. He first speculated that the gravitational force decreased in the same way but abandoned this in favor of a force that decreased linearly with distance. Note that he decided this on observational grounds. The speed of a planet varied during its revolution and was inversely linear with the planet's distance from the Sun. He thought this must be because of the change in gravitational force and thereby arrived at the inverse distance conclusion. There were no astronomical data or theories to support an inverse square law. This had to await the use of his three laws by Newton.

Kepler believed that God had created a harmonious, self-consistent universe in which, at some deep level, all things were simply connected, so he asked himself simple questions in an effort to find this harmony. Why were there only six planets?

Why did their distances from the Sun and their speeds have the particular values that were observed? Since there are infinitely many possibilities, there must be reasons why God had made these particular choices. Geometry was the mathematics of the time and the mainstay of astronomy, so Kepler searched for a geometric answer and found it! At first he tried to relate the orbits to plane figures and got nowhere. But space is three dimensional, so he looked at possible relations with solid figures and arrived at a striking conclusion. The number of planets and the diameter of their orbits were determined by the five regular solids: the tetrahedron, the cube, the octahedron, the dodecahedron, and the icosahedron, with four, six, eight, twelve, and twenty faces respectively! For each of these, the shape and area of all faces are equal, as are the lengths of every edge and the angles between any two faces. These are the regular, Platonic solids; they are the only regular, concave solids, and there can be no others.

Consider the orbits to be the equators of successive spheres. If the sphere around Mercury is circumscribed by an octahedron, that around Venus with an icosahedron, that around Earth a dodecahedron, that around Mars a tetrahedron, and that around Jupiter a cube, then the six spheres inscribed and circumscribed around the regular solids will contain the orbits of the six planets. When Kepler compared the ratios of the orbits so defined with data, he found agreement within about ten percent. He ascribed the discrepancy to errors in the Copernican tables and believed he had found one of the fundamental harmonies of the universe! This fueled his lust for Tycho Brahe's data, which he knew to be much more accurate than the results of Copernicus.

Kepler now thought he knew the Divine Logic behind the structure of the planetary system. The five regular solids, the most perfect expression of solid geometry, defined the orbits of the six planets, the only number allowed. He was a committed Copernican when he had this insight, and he took his geometric scheme to be another confirmation of the heliocentric system, which he found much more elegant than the Tychonic hybrid of geocentric and heliocentric models.

His belief in the unity of nature as an expression of the Will of God also led Kepler to a remarkable conclusion regarding harmonious musical intervals. The discovery of the relation of the lengths of vibrating strings to such intervals by the Pythagoreans was the reason for much of their number mysticism, and was taught in the schools right up to Kepler's time and beyond. New musical theory had added a few intervals to the four preferred by the Greeks, and Kepler concluded that these too were harmonious and must be included as defining musical harmony. The harmonious ratios of string lengths contained the numbers 1, 2, 3, 4, 5, 6, and 8, but *not* 7. Kepler wondered why 7 was omitted and believed he found the answer in plane geometry. The polygons with three, four, five, six, and eight sides could all be constructed with a straight edge and compass, the only tools allowed in "pure" geometry. A seven-sided figure could not be constructed using a ruler and compass alone. It was appropriate that any musical interval with a 7 in its ratio had to be dissonant. Here was another manifestation of harmony, unity, and simplicity in God's plan for the universe. Music was not just an ancillary interest. The ultimate unity of nature emanating from the Divine Will demanded that musical harmony and the planetary system be related. He thought of the planets as actually producing tones as they raced through the ether and supposed that their angular speeds, as viewed from the Sun, were governed by the intervals in a musical chord. On working out the orbital distances from this, he found them to be in better agreement with data than his regular solid construction. This direction of inquiry, along with a detailed analysis of the orbits of Mars and Earth, led him to his second law and then to his first law.

Think of it! Kepler was a man with a number of religiously inspired notions of how the world works, and they led to his final truths by a tortuous route that ultimately forced him to give up his cherished ideas. He had inherited and firmly believed the idea that celestial orbits had to be circular. Yet he came to the realization that planetary orbits were elliptical. He thought he had found a great, fundamental truth in the relationship of planetary orbits to the regular solids and again in their relation to musical intervals, but was forced to abandon these. He was driven by two basic forces: one was a belief in the unity of nature, as all reductionists believe to this day, and the other by a commitment to the authority of empirical data. He knew Tycho's data were accurate, and when they contradicted his models, he discarded the models but kept looking for unifying laws. The labor was immense. Fitting the data to models and finding the correct shapes of the orbits and the relations among their properties took many years of dedicated effort. His inner turmoil must have been extreme.

The attempts to tie astronomical theory to divine harmony and geometric perfection reflected Kepler's religious convictions, but they were an obstacle to the general acceptance and dissemination of his correct work on planetary orbits. His contemporaries and immediate successors were put off by his mysticism, which I will call "speculative mysticism", because it is based on preconceived notions having little to do with experimental facts. This is in stark contrast to the "scientific mysticism" in which one is led to the ultimate mysteries by rigorous observation, experiment, and logic. It took some time for the useless speculations to fall away and for the three laws of planetary motion to be fully appreciated. But for Kepler himself, there was no dichotomy in his search for understanding. His passionate search for deeper meanings was a unity. He wanted to find the underlying harmony in the motion of celestial bodies, and his search first focused on what he thought were legitimate possibilities reflecting the fundamental unity of God and nature. When the observational data showed that his ideas were wrong, he abandoned them and moved on. He was committed to the facts, and so his "speculative mysticism". gave way to "scientific mysticism". Yet there is no denying that Kepler wanted to learn *why*, not merely *how*, nature worked as it did. In this sense, he was not really a modern scientist.

3

The first modern giant

Galileo and Kepler were contemporaries, and together they fully confirmed the Copernican model. They were very different kinds of men, both personally and professionally, and had very different family backgrounds. Galileo's family could best be described as living in genteel poverty; his father Vincenzio was an accomplished musician and wrote a highly regarded book *Dialogue on Ancient and Modern Music* on musical theory. He tried to supplement his income as a wool merchant, but could never earn quite enough to support his family comfortably. His financial situation actually got worse as he got older because he became ever more interested in musical theory and paid decreasing attention to the wool trade. He had hopes that Galileo would become the family's financial savior and therefore steered him into the study of medicine, which was the best paid and most respected of the professions available to him.

Unlike Kepler, Galileo had a father who cared about him and in several ways was a model and an inspiration. Vincenzio was highly intelligent and an accomplished mathematician who taught Galileo to respect mathematics as well as to play the lute. With his study of musical theory as an example, he taught Galileo to think independently and to suspect pronouncements based on authority alone.

At the age of nineteen years, Galileo attended lectures on Euclid, which he found irresistible. Vincenzio might have regretted this because Galileo abandoned the study of medicine for mathematics. While Kepler was modest and unassuming, Galileo was an outgoing extrovert with a fine sense of humor, who enjoyed good friendship, good conversation, good food, and good wine. He was irrepressible and boisterous and always ready to loudly criticize those he thought to be wrong. In fact, as a student, he early earned the nickname "wrangler" because he was so consistently disputatious.

He was a natural philosopher (physicist) as well as a mathematician and took the study of all nature as his domain, while Kepler stuck primarily to astronomy.[17] Kepler had religious beliefs about how nature worked, while for Galileo, there was no source of truth about nature except nature itself. Galileo did not have to go through the painful transition from speculative, pre-existing notions about the basic harmony of the world to fact-based conclusions that marked Kepler.

Many educated people, including astronomers and natural philosophers, and especially churchmen, often cited tradition, ancient beliefs, professorial authority, and religion as reasons for accepting or rejecting descriptions of physical reality. Not Galileo: he insisted that only observation and experiment could determine what was true in nature. This fundamental tenet, along with the methodology he developed, made him the first of the modern scientists. His skillful construction of experiments, designed to get specific data and answer specific questions, his combining data and mathematical reasoning, and even his use of thought experiments were the beginning of the scientific method. His approach and methods are still those of science today.

As always, there were predecessors and a gradual growth of knowledge so no specific date can be assigned to the emergence of modern science. But Galileo's life work was so important, and the methods he used were so extensively and consistently applied, that he stands out as the greatest progenitor of modern science.

Galileo was born just as the Renaissance in art and literature was winding down and as the Counter-Reformation was gearing up so he benefited from the new freedom of thought and

[17] This is not entirely true. For example, Kepler did some important work on geometric optics, but this arose from his use of telescopes for astronomy.

expression, while running right into the unyielding protectionism of conservative religion.

It was not until 1589, at the age of twenty-five, that Galileo was able to get a job as a lecturer at the University of Pisa, with the help of influential friends and two lectures on the dimensions of Hell in Dante's *Inferno*. In 1588 he was invited to give these lectures by the Academy in Florence, which were the beginning of his public life. He was an engaging speaker with a charismatic personality, his calculations were presented in a manner that non-mathematicians found easy to understand, and the subject matter was fascinating. Florence revered and honored Dante as its greatest poet and as the prime example of Florentine pre-eminence in the arts. The size and shape of Dante's Hell was an important and serious issue that had been debated for many years, and there were two major positions in the current discussions. One view had been presented by a Florentine and the other by a Venetian. Galileo knew that these lectures were a kind of test and wisely chose to champion the Florentine against the foreigner. From an allusion in the *Inferno* to an existing statue, Galileo constructed a chain of logic and calculation that gave the dimensions he sought and even the size of Lucifer's body, which was frozen in ice! It was a tour de force. The lectures were attended by much of the city's literate public, and Galileo became a local celebrity. The lectures helped him get the position at the University of Pisa.

An interesting, and perhaps important, circumstance is that the lectures led Galileo to the correct study of the influence of scaling on the strength of materials. He had stated that the spherical cap over the cone that was Hell was thick enough to support its weight. But he neglected to take into account that strength does not scale with size. The bones of an elephant, for example, have to be much larger, with respect to body weight, than those of a housecat. If the cat's bones were scaled up in the same proportion as the rest of the cat's body, they would break under the elephant's weight. Galileo had in fact used this rule to show why large ships in dry dock would not support their own weight. Apparently, he soon realized his mistake, but did not mention or publish it until he wrote the

Dialogue on Two New Sciences when he was an old man under house arrest by the Inquisition. Scientific discussion at the time, as all intellectual discussion, consisted of sharp, often personal attacks, and general knowledge of his error would certainly have invited such attacks and damaged his reputation.

He stayed at Pisa only three years, when he left for a better paying job at the University of Padua. Even then, he had to enhance his income by private tutoring to fulfill his growing family responsibilities. He was an outstanding teacher, but he apparently did not like to lecture, preferring interactive discussions with small groups of students. Yet his lectures were said to be always packed. His income at Padua was inadequate[18] so he tutored private students to earn extra money, and the number of such students that boarded with him rose to twenty.

The University of Padua was a truly international institution. Students from all over Europe came to study, or at least to party, under the tutelage of a stellar constellation of professors, and it hosted many of the most distinguished people of the time. Harvey was in attendance at Padua while Galileo was there, and although it is doubtful if they ever met, Galileo came to know and admire Harvey's work on the circulation of the blood. Another sign of Galileo's broad interests was his study of William Gilbert's work *De Magnete*, which was published in 1600, and his repetition of many of Gilbert's experiments. By covering lodestones with iron, he was able to greatly increase their strength.[19]

Padua is only about 25 miles from Venice, which was the wealthiest and most progressive city-state in Europe, and full of entertaining attractions. Padua was under the rule of Venice and benefited from its democratic and tolerant governance. Although power was held by a number of noble families, it was obtained and

[18] His income was always inadequate. He had many family responsibilities, including several dowries he had to provide, and an irresponsible brother who was a source of expense, rather than help. Also, he liked to live well, and did not deny himself life's pleasures.

[19] The iron coat was called *armor*, from which the modern word *armature* is derived.

dispensed by elections rather than merely as a birthright. Galileo had little to fear from the tradition bound Aristotelians in such a climate, but he was homesick for Tuscany and he wanted more free time for himself. As he said, a democracy demands service for public support, so he had to teach and to respond to requests for service, all of which bit heavily into his time for research. The Grand Duchy of Tuscany, on the other hand, was ruled by one man and could grant Galileo the free time he wanted. Thus in 1610 he returned to Florence as Chief Mathematician of the University of Pisa and as Philosopher and Mathematician to the Grand Duke. With no teaching duties, he spent all his time on research.

It was a fateful decision. Venice was relatively independent of papal power and in fact had expelled the Jesuits over a dispute with Rome. Tuscany, on the other hand was closely engaged with Rome, the Medicis having provided many distinguished members to the Church hierarchy. Tuscany was committed to the conservatism of the Church; Venice was not. The conflict between Galileo and Rome might have turned out quite differently if Galileo had stayed at Padua.

Galileo knew of and admired both Brahe and Kepler, although he had little use for the overly mystic strain in Kepler's thinking. He had used the appearance of Tycho's nova as evidence against the Aristotelian concept of the immutability of the heavens, and, from his correspondence with Kepler, it was clear that Galileo was an avowed Copernican as early as 1597. But, surprisingly, the three laws of planetary motion had no effect on Galileo's work, even though Kepler had sent him a manuscript describing them in detail. It seems that Galileo read only the preface and never went into the body of the text. Several reasons have been advanced for this neglect, the first being that Kepler was a terrible writer. He wrote in a confusing, wordy Latin that required a lot of hard work to be understood, while Galileo was a master communicator, whose writings and lectures were masterpieces of clarity and style. It is easy to imagine him losing patience with Kepler's dense, obscure verbiage. The second is that, while Galileo thought Kepler was an excellent astronomer, he was contemptuous of the magical streak in his thought. Galileo wanted to appeal only to nature, not to supposed harmonies and geometric numerologies. So Galileo was ignorant of Kepler's crowning achievement. It is interesting to speculate on how far Galileo could have gone with that knowledge, especially since he had the command of motion and mechanics that Kepler lacked.

Galileo was one of the few scientists who acquired a reputation as a great, almost superhuman, genius among the general public during his lifetime. Only Newton and Einstein were similarly venerated by society at large. These three, more than anyone else, have come to personify the meaning of scientific genius. All three became widely admired figures because of their work on the cosmos. Einstein because of the observations on the bending of light by massive stars that confirmed his general theory of relativity, Newton because his law of universal gravitation ordered the movement of the planets, and Galileo because of his telescopic observations of the night sky. Newton's public persona faded somewhat with the passage of time, but Galileo's has not. He is still recognized as a titanic figure, not only in the scientific community, but also by a public that knows little of the inner workings of science.

The *Inferno* lectures helped him get a job and brought him some local celebrity, but it was the publication of his astronomical observations in a little book, *The Starry Messenger*, in 1610 that brought him widespread recognition. Galileo was an excellent writer of captivating prose, and his writings are still regarded as ranking with the best literature in the Italian language. His literary talent and his ability to present scientific material in simple terms, and his skill at self-promotion, gave his book a wide audience and made him famous throughout Europe.

His telescopic observations, and the notoriety of his famous conflict with the Church, killed the geocentric theory once and for all, not only for astronomers, but for all literate society. He was in his mid-forties when he heard about the telescope in Holland and built one for himself of higher quality and with much more power. He turned it on the night sky and saw that the Moon was not featureless, but had plains and mountains much like the Earth, thereby reinforcing rejection of the Aristotelian belief that celestial and terrestrial matter were intrinsically different. His observations of sunspots, which he used to show that the Sun was rotating with a period of 27 days, showed that the Aristotelians were wrong to believe that the heavens were perfect while Earth was corrupt. He was already a convinced Copernican, and the observations of Jupiter's moons and Saturn's rings (which he saw as two satellites almost touching the planet), and particularly the phases of Venus, which could only be explained if Venus orbited the Sun, gave him irrefutable observational proofs.

The telescope also showed him that the Milky Way was really an agglomeration of a huge array of stars, making the universe much bigger than anyone had imagined, in keeping with the absence of measurable stellar parallax in a Copernican system.

It was an amazing novelty. Seeing distant objects as if they were nearby seemed magical to most people. This, coupled with the fact that it revealed so much that was new and strange in the skies, and strongly supported the revolutionary idea that the Earth moved, was enough to capture the imagination and admiration of all Europe. Galileo had also heard of the Dutch invention of the microscope and built a much improved version for himself. He used it to impress some influential people, but it never became an instrument of radically new research for him. It was Robert Hooke who later showed that the microscope revealed new worlds of the small just as the telescope showed new worlds of the large.

It is hard to imagine the dramatic and far-reaching impact of Galileo's observations. Suddenly, it was shown that the heavens were not perfect: the Moon had an irregular surface just like the Earth, and even the Sun had blemishes in the form of sunspots. There seemed to be no difference between celestial and terrestrial matter, and the Earth was no longer the center of the universe. The enormous construct of astronomical and physical knowledge, so carefully honed for a millennium, and its beautiful integration with religion, so that nature, God, and doctrine meshed in complete harmony, was being savagely destroyed. The turmoil among astronomers, natural philosophers, and clerics was very great. Within an incredibly brief time, Galileo had upended all knowledge and seemed to be tearing it out by its roots. The forces of conservatism knew what was happening but took some time to organize their deadly opposition.

Galileo visited Rome in 1611 to demonstrate his telescope and was lionized by everyone, including the most influential people in the Church. He therefore believed he could publicly commit himself to the Copernican model, and he did so in a series of letters. He sought permission from the Church to publish his conclusions and spent three years in negotiation with the authorities.

The Copernican model was well known, but regarded by most as a mathematical method of calculation rather than as an expression of reality. The tradition of separating calculations from physical reality went back to the Greeks, and as long as this was accepted, the Church had no objection. Aristotelians believed that it was the qualitative nature of things that was important for a true physical understanding, not mathematics, which was needed only as a methodology for calculating celestial positions. For them, the question of the physical truth of a mathematical planetary model made no sense. Galileo himself used the Ptolemaic model in a course for medical students who studied astronomy so they could cast horoscopes in connection with their practice of medicine.

Through his words and writings, it was obvious that he thought the heliocentric solar system was a physical fact, not just a computational convenience. He knew that this was a sensitive subject. In fact, the relation between the Scriptural accounts of the cosmos and scientific astronomy was an important theological issue of the time. A literal reading of the Bible was totally at variance with the new astronomy and yet was favored by most clerical authorities. The Earth had to be the center of the universe because Christ was born on Earth, and, to give the concept of Redemption meaning, the heavens had to be perfect and the Earth corrupt. All Biblical references were to a stationary, not a moving Earth. After all, Joshua ordered the *Sun* to stand still, not the Earth. Galileo, on the other hand, said that the Scriptures should be reinterpreted in the light of scientific knowledge. He pressed his case and was finally told that his publication would be approved on condition that the heliocentric model was presented as a hypothesis convenient for calculation and not as a physical fact, a position that the Church had held all along. The separation of computational tools from physical reality characteristic of that time is worth stressing. In fact, mathematics and natural philosophy (science) were two separate and distinct professions, as evidenced by Galileo's insistence that he hold the title of Philosopher to the Grand Duke of Tuscany, as well as Chief Mathematician of the University of Pisa, when he left Padua for Tuscany.

Galileo finally published his *Dialogue Concerning the Two Chief Systems of the World* in 1632. Actually, it contained three main characters, not two: Salviati, the knowledgeable proponent of the new science; Sagredo, an intelligent colleague ready to be convinced of the truth; and Simplicio, a convinced traditionalist who had trouble following Galileo's logic.

But his disclaimer that the Copernican system was only a mathematical hypothesis was weak, and it was placed near the end of the book. Furthermore, it was put into the mouth of Simplicio, the Aristotelian dolt, rather than that of Salviati, the brilliant modern scientist. No one reading the book could doubt that Galileo was proposing the physical reality of the Copernican system. Furthermore, Pope Urban VIII believed that he himself was the model for the obtuse, tradition-bound Simplicio, because he was clearly espousing the Pope's views. And, as if in an act of further defiance, the *Dialogue* was beautifully written in the vernacular Italian, thereby making it generally accessible to the entire population, rather than in Latin, the language reserved for scholars.

It is interesting to note that the Censor who first approved publication of the *Dialogue* unwittingly performed a great service for Galileo and for the scientific community. The title Galileo originally chose was *On the Flux and Reflux of the Tides*. This was unacceptable to the Censor, Nicolo Riccardi, because the ebb and flow of the tides was Galileo's strong argument for the motion of the Earth. Galileo claimed that tidal motion demonstrated unequivocally that the Earth moved. The diurnal motion could only be understood if the Earth both rotated on its axis and revolved around the Sun. This thesis was widely known and widely discussed, so Galileo's title was no less than a blatant affirmation of the Copernican theory, so Riccardi changed the title to reflect a discussion of the two competing astronomical systems. Without realizing it, Riccardi had done Galileo a double favor, because Galileo's argument on the tides was wrong, and because the title change focused the book on what was really the important issue.

The opposition to Galileo went beyond the objections to Copernicanism. There was a sharp confrontational debate between those who believed in the results of the new scientific approach and the traditionalists who wanted to maintain the Aristotelian science and Ptolemaic astronomy with its comfortable integration into Christian doctrine. The debate took the form of letters and books as well as public discussions, which were often full of personal attacks. Sarcasm and insults were part of the mode of discussion of the times, and the new science could not win such a debate because the traditionalists held the religious and secular power. Galileo was the leading figure of those advocating the new science and easily made some serious enemies. His red hair was a good symbol of his personality. Feisty, argumentative, and satirical, he was an arrogant opponent, brilliant and caustic, who loudly exposed ignorance and stupidity, with no regard for position or sensibilities. Since he was usually right, his diatribes were as damaging as they were insulting. It was some Jesuits who had suffered by Galileo who were instrumental in bringing him before the Inquisition and who convinced Pope Urban VIII that he was the model for Simplicio. It was this conviction that turned the Pope from a friend into Galileo's vindictive and permanent enemy.

The Copernican cosmology had been officially declared a heresy in 1616, and the publication of Galileo's *Dialogue* less than twenty years later was taken to be a direct challenge to ecclesiastical authority. Only the Church could interpret the Scriptures, and heliocentricity contradicted so many Biblical passages that it had to be wrong! The Church could not afford to have such a famous and respected figure as Galileo defy its authority by disseminating a heresy and challenging its monopolistic right to interpret the Bible. Two points should be clarified. The first is that the "Galileo affair" was not merely a struggle between the Ptolemaic and Copernican systems of the heavens. The debate centered on these two theories had been public and vigorous ever since Copernicus' great work was published, with as many people siding with one as with the other. And many of the objectors were not astronomers or mathematicians; they were philosophers or humanists that opposed Copernicus on grounds of principle and philosophy. The Church opposed Galileo because the geocentric view was supported by Scripture and the Christian–Aristotelian synthesis.

The second point is that characterizing the trial of Galileo only as a personal battle with the Church is misleading. True, the confrontation was between Galileo, who was unable to conceal his conviction that heliocentrism was physically true, and a number of individuals who, for a complex of personal, political, and theological reasons, opposed him. There were, in fact, eminent church leaders who were sympathetic to Galileo and many who did not want him brought before the Inquisition. It is also true that a basic problem was that the Pope and others in the hierarchy took Galileo's work as a direct attack on their power, and the Pope was furious at, as he believed, being the subject of ridicule in the person of Galileo's Simplicio. Power was so centralized that whatever the Pope decreed became, without question, law. The Church still suffers from this central control today. But the Galileo affair runs deeper than these factors. The very legitimacy of the power of the Church was threatened. The Roman Catholic Church claimed to be the *only* source of ultimate truth, and its senior members to be the *only* proper interpreters of the Bible and the only group with the authority to assert and decide doctrine. The new science threatened to undermine this power. The clerical hierarchy was incensed by Galileo's insistence that the Bible had to be reinterpreted in the light of new scientific advances, and that he did not hesitate to advance his own reinterpretations. As a Divine institution, empowered by the word of God, and with an infallible leadership, a confrontation was inevitable. It was a war between divinely inspired authority and the results of experiment and reason.

The Church won the battle, but lost the war.

The "Galileo affair" is a stark example of how power is exercised by true believers. Logic doesn't matter, truth doesn't matter, and life doesn't matter; only God's Word matters. If God tells you what is right, there is no possible argument against Him, and anyone who defies His word is a deadly enemy. Galileo came close to being killed by the Inquisition, and he was actually threatened with torture and death. He was spared and given a relatively light punishment because he was such a famous world figure and because he had some friends in high places.

In his old age, after complex interactions and discussions, Galileo was tried, found guilty of disobeying an order of the Church, and "vehemently suspected of heresy". The Inquisition spent a lot of time preparing for the trial, and the trial itself lasted over five months. At last, on June 23, 1633, he was called for the last event of the trial.

The Pope himself presided at the trial. Galileo was made to recant, abjectly, completely, and publicly. He was made to profess his final confession of error in a large hall filled with cardinals, bishops, and priests. Broken and afraid, he said that he had always believed and will always believe, all that is taught by the Church, and that he would never say anything to the contrary.

The importance attached to the trial is evident from the fact that the Pope ordered Galileo's renunciation to be sent to every Catholic university and to all professors of mathematics or astronomy with orders to read and disseminate it. The intent was to keep scientists and astronomers from challenging ecclesiastical authority. The Church succeeded in accomplishing its objective.

Galileo was ordered to recant, to remain silent on the subject of Copernicus, and sentenced to house arrest at the residence of the Tuscan ambassador. After a time, he was permitted to return to his home outside Florence, where he could readily be in contact with his devoted daughter, a nun in the convent of San Mateo.

He remained under house arrest, forbidden to publish or to have any scientific discussions with anyone. His *Dialogue Concerning the Two Chief Systems of the World* was placed on the Index of Forbidden Books and not removed until 1835. This position may seem strange to the modern mind, considering that Aristotle was vehemently opposed by all Catholic clerics (on the basis that he was a pagan) until the great Scholastic synthesis by Thomas Aquinas. In fact, the "Galileo affair" was an embarrassment to the Catholic Church that was officially ignored until 1979, when Pope John Paul II appointed a commission to reconsider the Galileo case. In 1992, the Pope officially acknowledged Galileo's unjust treatment and admitted "errors were made".

It was a humiliating end to a remarkable life, and Galileo came to symbolize the conflict between progress and orthodoxy and between science and religion. Then and now, the trial by the Inquisition was a large part of the Galileo legend. It was a major factor in the decline of the authority of the Church and in the rise of modern science as the prime method of acquiring reliable knowledge. Old, in poor health, and gradually going blind, Galileo made no more public statements on the motion of the Earth. Instead, he consolidated his work on mechanics and strength of materials in *Dialogues Concerning Two New Sciences*, published, after many delays, in 1638.

The effect of the Galileo trial on Italian science was profound. Innovation was unwelcome and dangerous, so scientific leadership passed to other parts of Europe.

Galileo has been criticized by some writers for not publicly and unequivocally maintaining his belief in the geocentric picture of the solar system. They believe he should not have caved in to the Church but should have heroically stood his ground for truth and for the cause of science. But he was a sick old man who was faced by the most powerful enemies possible and his position was complicated by the fact that he thought of himself as a faithful Catholic. He was no ascetic reformer and had always enjoyed the good things of life, and now he was threatened with torture and death. The threats were explicit and real. He recanted to escape these horrors. At this remove, I am not willing to judge that he should have done otherwise.

Astronomy made him a public celebrity, but Galileo himself believed that his studies of motion and falling bodies were his most important contributions, and he was right. He was the first to examine motion and gravitation using sound scientific methods, thereby making all future progress possible.

He found that all bodies fell with the same velocity and acceleration when dropped from the same height, refuting Aristotle's contention that the speed of fall was proportional to the bodies' weight.

Others had concluded that all objects fall at the same rate, but they made little impact on the science of their day. Galileo had much greater influence because he was more famous and an articulate anti-Aristotelian. In an age when there were no scientific journals and organized scientific societies such as Accademia dei Lincei did not circulate scientific results very efficiently, he was a unique disseminator of his work. Books and papers were often written in manuscript form, and handwritten copies were sent to a select few. But Galileo carried on a large correspondence in which he described his results, and published printed books as well as manuscripts. And his discoveries were so dramatic that they captured the imagination, so they were widely discussed and talked about. His observations through the telescope spread through Europe in a matter of weeks, rather than years, and within five years after it was first published, the *Starry Messenger* appeared in China in a Chinese translation. Galileo was a world celebrity.

From experiments on balls rolling down inclined planes and the motion of pendulums, he correctly inferred the law of falling bodies, which states that the velocity is proportional to the time after being dropped, and that the distance is proportional to the square of the time. He found that the path of a projectile was a parabola. Most importantly for future theory, he formulated the principle of inertia, the idea of a force producing acceleration, and the relation between the observations in systems moving relative to each other with constant velocity.

These were momentous results, and not only because they showed that Aristotle was wrong. They were the first steps on the road to understanding gravitation and were a part of the work that made it possible for Newton to create his theory of gravitation. Galileo's observations that *all* bodies fell at the same rate, irrespective of size or constitution, and his concept of inertia were ultimately developed and refined to give the principle of equivalence, which is the foundation of Einstein's relativity.

The fallacy that heavier bodies fall faster than lighter ones is persistent. I had a German teacher in high school who firmly believed this to be true. I couldn't imagine how anyone could think such a thing. Even without doing the experiment, it was obvious that, except for the resistance of the air, a body would not fall half as fast as one twice its weight. Imagine dropping a body that contained a mechanism to cut it in two as it was falling. Could anyone believe that the speed of each piece would suddenly be reduced to half its value? Or, if two equal bodies were falling side by side and a spring suddenly forced them together so they were glued into one, is it rational to think the velocity would suddenly be doubled? I gave my teacher these arguments, which I learned later were similar to Galileo's own thought experiments,[20] but she was not convinced until I stood on her desk and dropped a large and a small book at the same time. She accepted the fact, but thought it was unnatural. It is a simple result, obvious to those of us born after Galileo, but it is a prime example of how the simplest ideas can have the most remarkable consequences. A cornerstone of Einstein's general theory of relativity is just this: all objects dropped from the same height fall to the Earth at the same accelerating rate, no matter what their mass.

I cannot resist telling you about a demonstration I learned in high school. The problem was to experimentally verify Galileo's

[20] I have met several people who, in their youth, had also come to this thought experiment without knowing of Galileo's work. It seems so obvious now that it is a wonder it was not immediately known to everybody. Others in Galileo's time, and even before, had similar thoughts.

law of freely falling bodies. The apparatus consisted of a long sheet of waxed paper pasted on a long, smooth board and set upright on a stand, with a double-railed track alongside the board. A tuning fork was put near the top of the board and allowed to fall down the track. A small stylus was attached to the fork, just touching the waxed paper so that, when the tuning fork was made to vibrate and dropped down the track, it made a wavy indentation on the paper. The waxed paper simply recorded the vibration of the fork as it fell. The crests on the paper went further apart as the fork fell. From the known vibration frequency and the spacing between the wave crests on the paper, it was a simple matter to measure the time and the distance of fall. Out came Galileo's law that the distance varied as the square of the time! This was so much easier than Galileo's convoluted method of guessing what the law should be from geometrical considerations and then experimenting with inclined planes. Galileo could have done such an experiment had he thought about it. All the requisite technology was available. I don't know the name of the brilliant individual who first thought of this experiment. A number of other demonstrations of free fall have been devised, including making the measurements with photocells. In my view, nothing has matched the simplicity and elegance of the tuning fork method.

As usual, there were precedents that anticipated Galileo's work. The thought experiment on the motion of bodies before and after they are separated or joined, for example, had occurred to others who used it to question Aristotle. It was too simple a notion to miss. There were even those who concluded that Aristotle's contention that the speed of fall of a body was proportional to its weight was dead wrong. And some of them supported this with actual experiments and observations of falling bodies. There is still some question about the famous experiment of dropping two balls from the Tower of Pisa. Dropping balls from towers did not originate with Galileo, but the dramatic story of inviting a large crowd to witness the event, and the refusal of Aristotelians to even look at it, was probably an exaggeration propagated by Viviani, the most admiring student of Galileo's old age and his first biographer. In fact, the Dutch physicist Stevin had performed such an experiment long before.

There had been much previous work advocating observation, experiment, and the use of mathematics, and many were questioning Aristotle's mechanics. But Galileo towers above all the others. Of course there were predecessors, and their work was surely important, but, as usual, history honors the scientist who brings it all together and adds new and critically important factors. It is not too much to call Galileo the first modern physicist. Until his time, philosophical reasoning played an important, and sometimes determining, part in scientific investigation. Galileo steadfastly and stubbornly adhered to a simple principle: nature is always right. No matter what philosophical principles were at stake, no matter what reasoning or authority proclaimed, the way to find scientific truth was to observe, experiment, and measure.

Above all, to measure. Before Galileo, the only real application of mathematics to science was in astronomy, and even there, mathematics was primarily a means for calculating planetary positions. Even the debate between proponents of the Copernican and Ptolemaic systems centered on their respective utility in computing. There was no real effort to use mathematics to investigate scientific fundamentals. Of course, this is not quite true. Archimedes used both measurement and mathematics in the modern sense of quantitative physics, especially in the study of hydrostatics and of simple machines, and Galileo was thoroughly familiar with his work. In fact Galileo admired Archimedes immensely and was inspired by his work in hydrostatics and mechanics. He knew the story of Archimedes' solution to the question of whether or not the King's golden crown was alloyed with a baser metal by displacing the crown in water; and he knew Archimedes' Principle, which states that a body in a liquid is subject to a buoyant force equal to the weight of the displaced liquid. Thus a solid would weigh less in a liquid than in air. Galileo was sure that Archimedes could not get accurate results, so he invented a sensitive balance for weighing objects under water.

But it was Aristotle's physics that dominated science and it was Aristotle's physics that Galileo replaced.

Aristotelian physics was the effort to find the ultimate causes of things, and these were always qualitative. Galileo shifted this old quest for "why" into the modern quest for "how". In the study of motion, for example, the old physics focused on where the moving body was and where it wanted to be. The passage of time was not a consideration. Galileo introduced time into his measurement and his theory, thereby converting the study of motion from the study of the "natural" tendencies of bodies to a study of dynamics[21] in the modern sense. With his combination of mathematics and experiment, and his process of testing a hypothesis by experiment, he created modern physics.

Modern science, in the sense of combining mathematics with measurement and experiment, of taking experiment as the final authority, and of concentrating on the observed phenomena themselves rather than looking for "essential natures" and teleological aspirations of matter, began with Galileo. In a sense, this was a greater achievement than his astronomy or his dynamics because it established a methodology that was useful for all studies of nature. He also perfected the necessary process of abstraction, by which a phenomenon was stripped to the basics, which may not be realizable in the real world, and putting in the real-world complications once those basics were understood. He created the scientific method and changed the nature of scientific investigation. After him the old ways were truly dead.

Galileo made Newton's synthesis possible by firmly verifying the heliocentric solar system and, more importantly, by his work on dynamics, which contained the beginnings of Newton's laws of motion.

He was not always right and was not always completely free of the old physics. Unlike Kepler, Galileo rejected the idea of

[21] Remember that kinetics is the study of motion, considering only position and time; dynamics is the theory of motion that includes the action of forces.

action-at-a-distance and did not think of gravity as a force acting between bodies. He believed it was an intrinsic property of matter, and he was committed to the idea that the most natural motion was circular, which did not require further explanation because it was preferred by nature. The concept of inertia of heavenly bodies for Galileo was that circular motion was preserved and could be changed only by applying an external force. However, when applying it to terrestrial mechanics, he used the concept in its modern sense, stating that a body persisted in a state of rest or uniform rectilinear motion unless acted on by a force. In his masterwork, *Two New Sciences*, he explicitly stated that a body moving at a constant velocity in a plane, not being acted on by any forces, would move with that constant velocity to infinity. This became Newton's first law of motion.

Galileo's treatment of circular motion was a great advance on several fronts. He was the first to treat circular motion as a compound motion and stated that if a body were to fly off on a tangent to the circle, the result would be rectilinear motion for which the law of inertia would hold. Galileo had discovered the principle of the vector sum of forces and velocities!

In spite of his commitment to observation and experiment, Galileo was not only an experimentalist; rather, he was the first modern theoretical physicist. The Copernican model was not really forced by experiment, and straightforward observation was against it. Galileo recognized that any number of different models could reproduce the quantitative data of his time and he chose Copernicanism because of its elegance and simplicity. In his *Dialogues* he actually used a model that was a little different in that he placed the Sun at the precise center of circular planetary orbits, rather than slightly away from the center, as Copernicus had done to get better agreement with observations of planetary positions. He probably would not have used Kepler's elliptical orbits even if he had known of them. His object was not to merely reproduce observational data, but to show convincingly that the Earth moved, and to do this he worked as a true theoretical physicist by using the simplest possible model.

Galileo's studies on mechanics bore the same marks. The concept of inertia was essential for the development of mechanics and for the connection between moving reference systems. Yet there were no experiments that demonstrated the inertia of moving bodies. Just the opposite was the case. Every moving body left to itself ultimately slowed down and stopped; none ever continued to move indefinitely. The swinging pendulum ultimately comes to rest and the billiard ball slows as it rolls over the cloth. It took profound insight to recognize that, as the opposing force becomes vanishingly small, there is no resistance to the indefinite continuation of uniform motion. I*n the absence of forces*, inertia is a fact of nature, but it was a thought experiment, not real data, which led Galileo to the concept of inertia.

A related thought experiment led him to the conclusion that phenomena in uniformly moving reference systems acted the same when referred to their own respective systems. He imagined a ship sailing along at a uniform velocity relative to the shore and concluded that, while an object dropped from the top of the mast displayed a curved path to an observer on the shore, it would appear to drop straight down to someone standing on the ship. Also, he concluded that if someone on the shore measured the velocity of a body moving uniformly on the ship, he would get a result that was the sum of the velocity of the body and that of the ship. Ergo: inertia and the principle of Galilean relativity. This was the first precise statement that nature behaves the same way at all times and places, and it is worth repeating:

The laws of mechanics are the same for all systems moving with respect to each other at constant velocity.

Note that this is the same principle that yields Einstein's theory of special relativity. The only difference between Galileo and Einstein is that, in transforming the formula for motion from one moving system to another (such as Galileo's ship and shore), Einstein took into account that light had a finite, constant velocity, while Galileo did not consider the velocity of light.

Even Galileo's work on the motion of falling bodies was that of a theorist. Apparently, he believed he could *deduce* the correct relation by logical reasoning. At first he concluded (incorrectly) that an object fell at a constant velocity, but later changed his mind and asserted (correctly) that it fell with constant acceleration. But this was not enough for him. Experimental demonstration was necessary. Time and distance could not be measured well enough for a freely falling body to get accurate results, so he resorted to measuring the time and distance for a ball rolling down an inclined plane. Interpretation of the data depended on two levels of abstraction. The first was that the resistance of the balls from the planes had to be neglected. Galileo greatly reduced the resistance by putting a very smooth, highly polished, regular groove in the plane, and constructing a very hard (brass) sphere as nearly perfect a sphere as he could manage. He could measure the distance the ball traveled down a plane much more accurately than for free fall. He used a water clock consisting of a vessel from which a small stream was released with laminar flow (so no turbulence would disturb the results), which was weighed to give a measure of the time.

The second abstraction was to identify the rolling ball with one in free fall by a geometric analysis. Abstraction was important in his entire approach to falling bodies and was often unexpectedly fruitful. Actual experiment did *not* show that when two bodies of widely different weights were simultaneously dropped from a height, they arrived at the ground *precisely at the same moment*. In fact there was a difference in their times of arrival. But Galileo noted that this distance was small and depended on both the size and density of the body. He correctly deduced the role of air resistance and considered the fall of bodies in media other than air, such as water. He concluded that the medium exerted a resistive force on a body moving in a fluid and that this force was proportional to the velocity. So his work on falling bodies was linked to an entirely different issue: the motion of bodies in fluids.

His interpretation of the experiments on falling bodies showed that he had a detailed and accurate idea of speed and acceleration that was totally absent from the thought of the Aristotelians. He computed the acceleration of a falling body as well as its velocity.

He found that the acceleration of a falling object on Earth was 9.8 meters per second per second. That is, every second, its speed increases by 9.8 meters per second, which is about 32 feet per second. Since the acceleration is constant, the speed increases linearly with time as the object falls, and the distance it travels increases as the square of the time. Thus, one second after a ball is dropped, it will have fallen 9.8 meters and acquired a speed of 9.8 meters per second. After two seconds, its speed will be $2 \times 9.8 = 18.6$ meters per second and it will have fallen $4.9 \times 2^2 = 39.2$ meters, and so on.

At bottom, a physical theory is just a guess. It may come from a thorough familiarity with the phenomena combined with a knowledge of all the mathematical and experimental tools involved and a sense of how nature works, but it is still a guess. If it organizes existing facts better than any previous theories in a convincing way and it suggests other experiments to gather new facts that also fit into the theory, then it works and is accepted. This is what Einstein meant when he said that science is a free creation of the human mind. Galileo's guesses were those of a genius and he was the first master theoretical physicist.

Furthermore, he was a reductionist in a deeper and more extensive form than Kepler, or any other previous scientist. Galileo was born into a scientific world in which there were two great dichotomies: one was the difference between celestial and terrestrial matter, which did not follow the same laws, and the other was the split between kinematics, which was quantitative and dealt only with the time–space relations of motion, and dynamics, which dealt with the causes of motion and was qualitative. Galileo's work united this four-way split.

Two New Sciences was his masterpiece. It cemented the use of mathematics in science by creating mathematical physics; it presented a definitive, organized description of motion; it anticipated much of Newton's mechanics; and it correctly and accurately described the action of the Earth's gravity. It was a tour de force that has been hailed as being second only to Newton's *Principia* in importance for the development of science.

4

The grid

Galileo's applications of quantitative theory to physics showed that he was a master of the mathematics of his time. His mathematics, however, was limited and inadequate compared to that available to modern scientists. Succeeding centuries could not have advanced without much more powerful mathematics. It was a mark of the genius of the seventeenth century that it was able to go so far with such limited tools, extending and modifying them as needed for specific problems.

The era of modern mathematics started with the creation of analytic geometry, which united geometry and algebra, and is generally credited to Rene Descartes, the most influential philosopher of the seventeenth century. As always, there were predecessors, and the association of numbers and geometric points went back to the ancient Greeks. More recently, for example, Galileo combined geometry and algebra in working out the parabolic path of a projectile. Also, Descartes' contemporary Pierre de Fermat independently worked out what amounted to analytic geometry in his study of maxima, minima, and tangents. Analytic geometry is just a way of associating a geometrical point with numbers. It is a simple idea with far-reaching consequences.

The connection between numbers and points on a straight line is obvious. Just mark some point on the line, label it zero, and call it the origin. Then any number can be thought of as a point on the line. To be explicit, mark off other points that are equal distances apart along the line and label them as integers. Negative integers correspond to equally spaced points to the left of the origin and positive integers correspond to points to the right of the line. Then every point is equivalent to a number. For example, the point midway between the marks labeled 2 and 3 has a distance from the origin of 2.5. Every other number is also equivalent to some point on the line and every point is equivalent to some number. A straight line is a one-dimensional coordinate system because it is a system of points in which every point is coordinated to a number. This was not new, even though many of the ancient Greeks did not like the idea, especially when the Pythagoreans found that assigning numbers to the sides of triangles sometimes produced numbers that could *not* be represented as the ratio of two integers. Since such numbers didn't seem to make sense, they became known as "irrational". While they make perfect sense to modern mathematicians, they are still called irrational numbers.

The points in a plane can be treated similarly, but two numbers are required. In hanging a picture on a wall, the height from the ground is not enough to specify where we should drive the nail; we also need to know how far we want it to be from the edge of the wall. That is, in a plane, we identify each point with two numbers because that is what it takes to specify the location of the point. To do this, we construct two numbered lines perpendicular to each other, one parallel to the bottom of the paper (usually called the x-axis) and one perpendicular to the bottom of the page (usually called the y-axis). To aid in locating points, lines parallel to each of the two axes are drawn through equally spaced marks on the axes, each mark representing a unit of distance. The result is a grid of little squares.

The representation of a point in a plane by *two* numbers, one along a line and the other along another line perpendicular to the first, was a revolutionary thought, because then every curve could be represented by an algebraic equation and every equation by a curve. The power of both algebra and geometry was thereby greatly increased because geometric problems could now be solved by algebraic methods and vice versa. Locating points in a plane by using two numbers was the foundation of the advanced mathematics so necessary for the growth of science. This is called a two-dimensional coordinate system because it is a system of points each of which is coordinated to a pair of numbers. These simple ideas are the foundation of the concept of the continuum, which is so important in physical theories. Later, we will meet the four-dimensional space-time continuum, which is essential in the relativistic theory of gravity.

Analytic geometry was a necessary precursor for the invention of the calculus by Newton and Leibniz.

Actually it was Fermat who explicitly defined a two-dimensional coordinate system like the grid we use today. But Descartes' *Geometrie* was the acknowledged masterpiece, not only because he was the most famous philosopher in Europe, but also because it laid out the entire theory of analytic geometry and its applications in a complete, coherent manner suitable for future advances. Also he published it for public consumption, whereas Fermat disclosed his most important results in the form of letters. But the association of points and numbers is quite clear in Descartes's work, even though he did not explicitly describe the coordinate plane.

Descartes's analytic geometry was the beginning of the modern method of defining mathematical spaces and led inexorably to the development of the mathematical theory, called differential geometry, which was essential for the development of relativity. It would have been impossible to work out general relativity without the tools of differential geometry.

Newton's connection to Descartes went beyond this. Like Newton, Descartes also was egotistical and selfish without much human warmth, although not as vindictive or as desirous of power. He also believed he had a messianic destiny and claimed to have had prophetic dreams. Three dreams on November 10, 1618 convinced him he would found a universal science and philosophy and that it was his duty to bring it to the world. With respect to science, this culminated in his *Principia Philosophiae* of 1644. It was enthusiastically received throughout Europe because it described a unified system that claimed to explain all of physics and astronomy, while incorporating the latest scientific results. The scientific revolution was in the air, the old ideas of Aristotle and the Scholastics were crumbling, and the need for a comprehensive replacement was urgent. Another reason for his popularity was Descartes's public opposition to the prevailing skepticism that science could not be certain and that it rested on probabilities. Descartes strongly denounced this, stating that only knowledge based on certainty was of any value and that he knew how to acquire this certainty by starting from a position that doubted everything except the bare minimum that was absolutely sure. This was the essence of his famous dictum "*cogito ergo sum*", by which he claimed to start with the extreme skeptical position that only his own existence could not be denied and then moving to certainty through mathematically precise logic. His method seems clear to us now; do not take anything for granted, start with the bare minimum that must be accepted, and on this basis, accept only what can be proved. But at the time, he started a revolution in critical thought about philosophical issues and has been called the father of modern philosophy. While this may be an exaggeration, given the teachings of the ancient Greeks, his work was nevertheless important because it brought into question many concepts that were being taken for granted. At the same time, he asserted the existence of objective truth and claimed that human beings could find it.

Descartes was warmly welcomed by the more progressive intellectuals, and his remarkable philosophies and novelties about the physical world were widely discussed by many who were ill equipped to understand him. But many Protectors of the Faith opposed him. His rejection of authority, other than his own reason, threatened the authority of revealed religion, and his completely mechanistic physics demoted God's position to nothing more than a First Cause with no further role in the universe. Descartes understood this opposition, and, in 1634 when he learned of the Church's condemnation of Galileo, he suppressed publication of his book *Le Monde*, a totally Copernican work.

Descartes's contentions of the unity of science, its mechanical foundation, and its necessary connection to mathematics were in accord with the modern wave of advancing scientific knowledge, but the details of his physics and astronomy were hopelessly wrong. He created these by pure reason, giving only minimal consideration to experiment or observation. Of his ten laws of motion, only two had any merit. They were essentially nothing less than Newton's first two laws and contained a precise idea of linear inertial motion. Descartes took linear inertia as being fundamental.

He agreed with other philosophers of his time that there could be no action-at-a-distance. Thinking that gravity, or any other force, emanating from one body could act over a distance on another body through completely empty space was patently absurd, so there had to be something in space to transmit the force. This was the ether, a real if highly rarified form of matter carrying all forces among material bodies through its motion. Descartes elaborated this ancient concept to produce a detailed mechanism for the solar system and the action of gravity. The Sun was proposed to be at the center of a huge whirlpool in the ether whose swirling motion carried the planets in their orbits. The planets were the center of their own whirlpools whose vortices carried their satellites. The Cartesian concept of gravitation retained the ancient idea that force could only be the result of one body acting directly on another. The ether did this for the gravitational force. Ether particles in the vortices collided with each other and then with solid bodies, so gravity arose from mechanical collisions and had nothing to do with action-at-a-distance.

The wide admiration for Descartes's *Principia* surely irritated Newton because he knew it was wrong. In his own *Principia*, Newton showed how the theory of vortices was inconsistent with Kepler's laws of planetary motion, with the basic laws of mechanics, and with Descartes's own assumptions. Most historians believe that the title of Newton's own great work was explicitly chosen to replace Descartes's with his own system, and substitute his own fame for that of Descartes. He succeeded.

Descartes's work was the last serious attempt to start with philosophy and use it to derive a science. After Newton, it was necessary to make philosophy conform to science, not vice versa.

5

The universal force

The three men sitting at a table in a London coffee house were deeply engrossed in an animated discussion that very few besides themselves could understand. It was wintertime and the hot coffee, warm fire, and heated conversation was a pleasant and fruitful interlude on that January day of 1684. One of the men was tall and relatively handsome, one was small with a large head, and the third was a little taller, but dwarfish, with a stoop and a somewhat crooked body. They looked very different, but their intellectual lives had much in common, and they could talk to each other on a level that was profoundly intimate yet non-personal.

Each of them has become a historic figure, remembered for a singular achievement: Edmond Halley for the comet that bears his name, Robert Hooke for his discovery of the basic law of elasticity, and Christopher Wren as the architect who designed St. Paul's. Their accomplishments and their importance in the history of science go far beyond these individual achievements, however. They were each brilliant with wide-ranging interests, and they were leaders of the scientific ferment that was putting England in the vanguard of the great movement that was creating modern science. All of them were astronomers and they were all recognized for their superior abilities.

Robert Hooke was the most versatile of the three. He was a talented artist and musician and actually started on a career as a painter when he was young. But it was in science, instrumentation, and engineering that he truly excelled. His most widely known work was in microscopy. Using a microscope he built himself, he examined the world of the very small and, for the first time, displayed the anatomy of insects, the cellular structure of plants, and the structure of everyday objects, such as the point of a needle, which, under magnification, was blunt and lumpy, or the magnified hairy mesh of textiles. Hooke drew the illustrations himself, using his artistic talent to accurately depict the wonders of the microscopic world. The book was written in a literate prose style, which, along with the astonishing drawings, guaranteed a wide audience among scientists and laymen alike. Its impact was sensational, especially the pictures of molds and insects, and showed that the world of the very small held wonders comparable to those in the recently discovered realm of the very large. In addition, the *Micrographia* described a model for the nature of light.

Hooke's optical studies put him in the great stream of progress leading to modern scientific knowledge. His experiments and speculations on the wave theory of light were adopted, refined, and developed by Huygens, and later by Fresnel and Young, to become the explanation for all of optics. It was completely successful until the discovery of quantum phenomena.

The bitter feud between Hooke and Newton started with optics because Hooke maintained that white light was a single unitary thing and that color was produced by the deflection of wave fronts at prismatic surfaces. Newton's work, which held that the colors of the rainbow were primary, and that white light was a mixture, was in direct opposition to this. Furthermore, Newton's corpuscular theory of light contradicted Hooke's wave theory, so the battle lines were drawn clearly and early. Actually, Newton offered the corpuscular hypothesis as only one possibility among several, since he preferred not to speculate, but Hooke was committed to the wave theory and saw that Newton was not. While the wave theory won out, Hooke's theory of color was completely replaced by that of Newton.

The microscopic and optical studies were Hooke's most famous, but by no means his only, works. He studied combustion and its relation to the components of air, and built and used vacuum pumps while assisting Robert Boyle in his work on the physics of gases. He was a superb engineer, as shown by his spring-driven escapement mechanism that was in common use until the advent of the quartz-regulated watch, and by his universal joint, a device essential for the modern automobile. He was intensely interested in flight, having studied both bees and moths and their differences of propulsion, and designed a variety of flying machines, none of which were ever built.

The City of London was rebuilding itself after the great fire of 1666, and Hooke was appointed Chief Surveyor, working with Wren to lay out a reconstruction plan and to design most of the public structures.

Through all of his activities, Hooke maintained an intense interest in astronomy, building telescopes, making careful observations, especially of planetary surfaces and comets, and working on theories of the solar system.

Christopher Wren was a mathematician and astronomer before he became England's premier architect. His interest in astronomy started with lunar observations in 1655; he was Professor of Astronomy at Gresham College and was appointed Savilian Professor at Oxford in 1661. His mathematical talents were formidable, and he did important work on the cycloid, geometric progressions, and logarithmic spirals. Newton himself, who seldom had a good word to say about the abilities of others, ranked him among the best mathematicians of his time. Like other virtuosi, his interests were broad, including anatomy, optics, and elasticity, as well as astronomy. His life changed when he was made a Commissioner for Rebuilding the City of London in 1666. He threw himself into this task with great energy and, with Hooke's able assistance, created a new London with more spacious streets and beautiful public buildings and churches.

The third man at the coffee house table, Edmond Halley, was the one most committed to astronomy, although his scientific work included meteorology, the creation of the first mortality tables relating death and age (for the city of Breslau), the relation between barometric pressure and elevation, and construction of a diving bell. He was already an accomplished astronomer by the age of seventeen and became an assistant to John Flamsteed, the Astronomer Royal, when he was nineteen years old. His important astronomical work included mapping the heavens of the southern hemisphere, studies of eclipses, and computing the proper motion of the stars since the time of the ancient Greeks. But it was the comets that made him forever famous. Using Newton's law of gravitation, he determined their paths and concluded that a number of past sightings were all of the same comet, the one observed in 1680 (that bears his name), and predicted it would return in 1758. He was intensely interested in the application of gravity to planetary motion and was after a decisive proof that the force exerted by the Sun decreased as the square of the distance between the Sun and the planets.

Seated around the coffee shop table, the conversation turned to the workings of the solar system, and Halley asked if not all facets of planetary motion might be derived from an attractive force decreasing as the square of the distance from the Sun. With some amusement, Hooke and Wren informed Halley that they already thought this to be the case. Wren noted, however, that while it was easy to guess at this, it was quite difficult to demonstrate, and a proof of the relation of the inverse square law to the solar system did not exist. Hooke insisted that he had indeed constructed such a proof but would not publish it because he wanted others to try and fail, thereby showing its, and his, true value when he finally made it public. But Wren was a much better mathematician than Hooke. This, and Hooke's well-known propensity to exaggerate his own accomplishments, prompted Wren to issue a challenge declaring that if either of them would supply a convincing proof within the next two months, he would give him a book worth forty shillings. None of the three was able to come up with such a demonstration during the allotted time.

Halley then decided to seek an answer from the only man he thought capable of finding it, so he visited Newton at Cambridge that spring, thereby starting a chain of events leading to the writing and publishing of the greatest scientific tract ever written, the *Philosophia Naturalis Principia Mathematica* (Mathematical Principles of Natural Philosophy).

Halley asked Newton what the orbit would be if the Sun attracted a planet with an inverse square law. Newton had already shown that an elliptical orbit implied the inverse square law and immediately replied that it would be an ellipse, that he had proved this long ago, and that he had mislaid the proof, but would reproduce it and send it to Halley. Months passed before Newton responded, but Halley just waited. His patience was rewarded by a manuscript that went far beyond the single answer he had asked for. Newton's paper contained the fundamental theoretical mechanics of motion and showed not only how an inverse square law of gravitation resulted in an elliptical orbit, but gave Kepler's other two laws as well.[22] With Halley's encouragement, Newton continued with the work that resulted in the publication of the *Principia* in 1687.

The depth and range of this work is astonishing. To properly address the solar system, Newton first created modern mechanics, defining time, space, mass, momentum, and force and thoroughly worked out the physics of moving bodies in a vacuum and in resistive fluids. He proved that the force of gravity was universal from a comparison of its effect on the Moon and on a falling body on Earth. He correctly computed its effects on all planetary orbits, comets, and the Moon, accounting not only for Kepler's laws and the gross orbits, but also for detailed effects such as orbital precession and perturbations. He computed the masses of the Sun and of the planets, showed that the tides are the result of gravitational pull from the Sun and the Moon and accounted for the oblate

[22] More generally, Newton's theory predicts that any orbit resulting from an inverse square law of attraction must be a conic section. That is, the orbit can be an ellipse, a parabola, or one branch of a hyperbola. Specific properties of the planets make this conic section an ellipse. An interesting point is that the second law does not depend on the inverse square law. It depends only on the fact that the force is directed to the center of the attracting body. This is related to the conservation of angular momentum.

shape of the Earth. Wave theory, elasticity, sound, the pendulum, projectiles, fluid flow: all were subject to precise mathematical analysis. He developed the mathematics of vortices, showing in the process that Cartesian physics was nonsense. Beyond this, he was the first to create a unified quantitative theory of nature by bringing together the laws of mechanics, universal gravitation, and celestial dynamics.

The *Principia* contains all this and much more, and yet the method he adopted is even more impressive. He sought coherent and consistent proofs throughout his work and demanded that any testable mathematical results agree with actual observation. Only Newton would have first created rational mechanics, with his three laws of motion and their application to bodies moving along general curved paths, before working on the theory of the solar system. And he did all this using geometry! One must try to go through some of the more complex geometric proofs to see how difficult this is. The use of analysis, in particular calculus, would have made it all so much easier! Newton surely already possessed many of the required methods of calculus, but the authoritative mathematics of the time was geometry and all attempts at serious mathematical physics were cast in the language of geometry. There is some evidence that he first used calculus to derive at least some results and then translated them into geometry. This would make sense but is not necessary because Newton was a complete master of all of geometry.

Newton combined a demand for rigorous mathematical proof with total obedience to the discipline of experimental data. Others had thought of universal gravitation and assumed it acted with a force that decreases as the square of the distance, and some, including Halley, had used Kepler's third law to show that this was indeed the case. Newton had already done all this during the plague years but would not even reveal his results, let alone publish them, because he had tried to *prove* that gravitation was universal and, by his standards, had not done so. If gravitation is universal, then the rate of fall of an object on Earth must be related to the revolution of the Moon around the Earth. The relation can be found from values for the acceleration due to gravity at the Earth's surface and the distance from the Earth to the Moon. When he did the calculation, he found an approximate agreement but not close enough to satisfy him. Years later, he repeated the calculation with better values for the Earth–Moon distance, and the agreement was then excellent. There was no doubt that Newton believed in universal gravitation, but now he could accept it as proven.

With hindsight, the calculation showing that the universal law of gravity extended from the Earth to the Moon is quite simple. Galileo had shown that the path of a projectile was the result of the force of gravity acting downward and the inertial tendency to move forward with the projectile's initial velocity. The compound motion described a parabola, which would eventually hit the ground. If the projectile's speed were high enough, and if it were shot from a high enough tower, the projectile would travel far enough that, because of the Earth's curvature, it would miss the Earth altogether and just continue going around it. Similarly, if the Earth's gravity is acting on the Moon, then its orbit is a compound of two motions, one parallel to and the other perpendicular to, a line joining the centers of the Moon and the Earth.

(It is easy to calculate the distance the Moon falls to the Earth in a given time. Just draw an extended line tangent to the orbit and calculate the distance from this line to the actual orbit. Note that although the Moon has moved down during that time, it has also moved perpendicular to the Earth's radius, so it never hits the Earth.)

The orbit of the Moon is the result of the balance between the gravitational pull of the Earth and the centripetal force arising from the Moon's circular motion. Newton knew the size of the orbit, the velocity of the Moon, and the fact that its distance from the Earth was close to 60 Earth radii. If the universal inverse square law were true, the gravitational force on an object at the surface of the Earth would be $60 \times 60 = 3600$ more than that at the Moon.[23]

Newton did the calculation and found that it agreed with the observed motion of the Moon.

A factor that disturbed Newton was that all calculations, including his own, were performed as if the gravitational forces emanated from a point at the centers of the Sun and the planets. But, if gravitation is universal, it applies to every particle of matter, and, since the force on a body is the sum of the forces for every particle of the Earth, there was no guarantee that the force of attraction on a body outside the Earth should be precisely towards its center.

Similar arguments hold for the other planets. Newton was not content until he was able to mathematically prove that a sphere attracted a body external to it *as if* all the sphere's mass were indeed concentrated at its center. Others were happy to state that the planets were so far from the Sun that they could be treated as point masseses to a good approximation. Newton demanded mathematical precision.

Once Newton was in possession of this fact, and once he had verified the validity of the inverse square law, the theory of universal gravitation was complete.

The content, structure, and methodology of the *Principia* are unique. It was such a comprehensive analysis of physics and astronomy, with such an overwhelming combination of experimental and mathematical proof, that any issues of priority of specific ideas are irrelevant. No one else has ever achieved such mastery over nature's secrets.

My interest here is in the law of gravitation, which is so simple and yet has such remarkable consequences. It states that two bodies attract each other with a force that is proportional to the product

[23] Note that this is the force on a unit of mass and does not give the relative weights of objects at the surface of the Earth and the surface of the Moon. Weight is the result of the attractive force at the surface and depends on the distance from the surface to the center as well as on the mass. The weight of an object on the surface of the Moon is about one sixth the weight it would have on the surface of the Earth.

of their masses and decreases as the square of the distance between them. No extensive mathematical facility is needed to understand this law: it is too easy. Imagine two people standing on the Earth. If one has twice the mass of the other, he will be attracted by twice the force. That is, he will weigh twice as much. Now think of the planets revolving around the Sun. The heavier the planet and the closer it is to the Sun, the greater the force of attraction. The law is much more specific than this: it says that if the distance between two objects is cut in half, the force between them goes up four times, whereas if the distance is doubled, the force decreases to one fourth its original value. Both masses, that of the Sun and that of the planet, appear in the formula because the law is universal. Not only does the Sun pull on the planet, but the planet also pulls on the Sun.

It even looks simple when written out as an equation:

$$F = -G\frac{m_1 m_2}{R^2}$$

Here, m_1 and m_2 are the masses of two bodies, labeled 1 and 2, R is the distance between them, and F is the force they exert on each other. G is a constant that is always and everywhere the same. This formula has been verified to a high degree of accuracy by observations of celestial bodies. Its universality was directly verified in 1798 by Henry Cavendish, who used a torsion balance to directly measure the force of attraction between two solid bodies in a terrestrial laboratory.[24]

It is a simple formula but it has remarkable consequences. First, the force of gravitation is always attractive. There are no negative masses, and gravity causes material objects always to attract and never to repel each other. For electricity, the force between two charges is also an inverse square law similar to that for gravitation, but there are positive and negative charges; unlike charges attract while like charges repel each other. Gravity, however, has only one sign.

[24] The formula reflects the usual convention that an attractive force is labeled negative.

Also, the gravitational force for moving masses is very different than the forces for moving charges. A moving charge induces a magnetic field. A moving mass does not induce any other kind of a force. Furthermore, the force between charges is not a simple inverse square when the charges are moving. Additional forces come into play that depend on the velocity of the charges. The law of gravity stays the same when the masses are moving, and has no velocity-dependent terms.

Note that the force increases very rapidly as the two bodies get closer together. If the separation between them is cut in half, they attract each other with a force four times as great. In fact the force quickly goes to infinity as the distance approaches zero. The coming together of the two bodies can be stopped only by some force large enough to resist the gravitational attraction. This has important consequences for the shape of the planets, for the sizes of celestial objects, for their internal structures, and for the very beginnings of the universe. The other point to notice is that the force decreases as the two bodies are separated, *but it never vanishes.* No matter how far apart they are, some force exists. The force extends forever, and that is why gravity is the glue that holds the universe together.

Whether the two attracting bodies are two apples, an apple and the Earth, the Earth and the Moon, a planet and the Sun, or two stars makes no difference. The gravitational constant is the same for all situations. It does not matter what the attracting bodies are made of. Apple, steel, water, or rocks, the gravitational attraction depends *only* on the masses of the attracting bodies and the distance between them. Furthermore, *there is no way to screen out the gravitational force.* Both the electrical force between charged bodies and the force between magnets can be screened by appropriate metal cages, but there is no way to nullify the force of gravitation. It is truly universal: always and everywhere present and always the same for *all* matter.

Gravity is a constant and pervasive part of daily life, and we live with it at an unconscious level as the unchanging factor in our environment, not giving much thought to how it shapes our existence.

Its profound significance can be clearly seen by applying Newton's law to the Earth itself. Let's start by asking a simple question. Why is planet Earth a sphere? After all, there are an infinite number of solid shapes, such as cubes, cylinders, cones, and tetrahedra. Why should the shape of the Earth have any regularity at all? The ancients had a straightforward answer, which was that the sphere was the most perfect geometric form, so that the Earth, along with all the celestial bodies, must be spherical. This was no answer at all, and no real answer was possible until Newton discovered the law of universal gravitation.

If any two bits of matter attract each other along a line joining their centers, than any large aggregate of matter must be spherical. To see this, consider two mutually attracting small clods. Ultimately, they get as close together as they can until the repulsive forces of their atoms prevent further motion. Now put a third clod into the picture. It will move towards the other two until it too gets as close as possible. That is, because of gravitation, an array of material particles will arrange themselves to make the distances between all their atoms as short as they can be, and, since the sphere is the shape that permits closest approach for all of them, they will coalesce into a sphere. The size of the sphere is determined by the balance of the force of gravitational attraction and the opposing force of the atoms, which repel each other more and more strongly as the atoms get closer together.

Incidentally, everyone knows a lot more about the forces between atoms and molecules than they consciously realize. We know this from ordinary everyday experience. Just think about water vapor. If the temperature is lowered enough, the water vapor condenses to liquid water, just as it does on the windshield of your car in cold weather. But this means that the molecules must be attracting each other. At high temperatures, the molecules are moving rapidly and they have a high energy. The energy of motion is called kinetic energy and is proportional to the temperature. The high speeds of molecules in hot objects keep them apart because the forces of attraction cannot overcome their kinetic energy. When the vapor is cooled, heat energy is taken away, the molecules slow down, and the attraction between molecules is strong enough to bring them together to condense into a liquid. If they are cooled further, the liquid turns to ice. All matter acts this way.[25]

Note that, once they are condensed, it is very hard to make the atoms come any closer. In fact, it takes very high pressures to squeeze molecules closer together once they form a condensed phase. The molecules must therefore be repelling each other very strongly. This gives us the following simple picture of intermolecular forces. There is both an attraction and a repulsion between molecules. When they are far apart, they attract each other, and the attraction increases as the distance between them decreases; but when they get very close, the repulsion completely overpowers the attraction. At a certain distance of separation, the attractive force and the repulsive force balance each other out. This distance is the equilibrium separation between the molecules, which is the actual separation when there is no molecular heat motion.

The size of the Earth is determined by the balance between the gravitational attraction among its parts and the force of repulsion among its atoms. The atomic forces are essentially electrical in nature, and it is interesting to note that they are very strong compared to gravitational forces. It takes 4.35×10^{42} grams of electrons to exert a gravitational force equal to the electrical force exerted by one gram of electrons. This means that it would take fifteen thousand Suns to generate a gravitational force equal to the electrical force exerted by one pound of electrons. However, practically all matter consists of both positive and negative particles that almost perfectly balance each other, so the interatomic forces are almost zero until the atoms get close enough. The electrons on two different atoms have the same sign of charge, so they repel each with a force that becomes very strong as the atoms get

[25] Superfluid helium is an interesting material in that it does not become solid, no matter how low the temperature, unless it is under high pressure. Below pressures of about thirty atmospheres, it remains a fluid even at temperatures very close to absolute zero.

close together. Of course, the electrons on one atom are attracted by the nucleus of the other atom, but the distance between the electrons on one atom from the nucleus on the other atom is much larger than the distance between the electrons on the two different atoms. The attraction is therefore much weaker than the repulsion, which can be strong enough to balance the gravitational attraction

Gravitation is essential for an understanding of many other properties of our planet besides its shape. The inner structure of the Earth, with its mantle, liquid portion, and solid center, is determined by the action of gravity;[26] the twelve-hour cycle of the tides is a result of lunar gravitation acting on the Earth's oceans; and the very existence of our atmosphere depends on the fact that gravity prevents the air from shooting off into space. Mercury, the Moon, and Pluto have no atmospheres because they are too small to hang on to any gases. Even the nature of our space program is dictated by the fact that a body must attain a velocity of 7 miles per second to escape the Earth's gravitational field.

Gravitation holds the universe together and indeed is the crucial factor in its formation. The birth of stars begins with the coalescence of atoms and dust and continues with the remarkable transformations of matter under the influence of high pressure. The changes from inert dust to balls of gas burning nuclear fuel, to the dense spheres and black holes that are the ashes of stellar evolution are all in response to the intense gravitational force existing in massive bodies.

The inverse square law has an air of "rightness" about it and had been surmised before Newton's masterly proofs. The astronomers that thought about the forces between the celestial bodies found it natural that they should decrease inversely as the square of the distance. The force between two electric charges, or between two magnetic poles, also decreases as the square of the distance

[26] The inner temperature of the Earth is the result of natural radioactivity in addition to gravitational collapse. But the radioactivity comes from nucleosynthesis in stars, which was the result of gravitation, so ultimately, gravitation is the origin of the Earth's structure.

between them, and this was also surmised before definitive proof existed. There is a simple reason for this expectation, as explained by Kepler's logic for the decrease of the intensity of light from a point source. Another compelling analogy is that of a spherical balloon being blown up, so its size increases. As it expands, the rubber skin stretches and its density goes down. That is, there is less rubber per square inch of surface as the balloon grows, *but the total amount of rubber stays the same.* The area of the sphere is proportional to the square of its radius, so the amount of rubber per square inch is proportional to the inverse of the square of the radius. This is a general rule; if something is being emitted in all directions, it must decrease by the inverse square of the distance from the source, simply as the result of elementary geometry. Even the ancient Greeks knew this since Archimedes had found the formulae for both the area and the volume of a sphere by a limiting process that was akin to the modern methods of the calculus. The ancient heliocentric model that was a main competitor to the Ptolemaic system included speculations on the inverse square law, which was justified by assuming that the gravitational force emanated from massive bodies' along rays perpendicular to the bodies' surface.

Thus the claim has been made that Newton did not discover the inverse square law, but rather adopted it from his readings of the ancient Greeks. It is a ridiculous claim and is in the same class as that of Hooke, who claimed credit for discovering the law of gravitation. Of course Hooke believed in the inverse square law, and so did some Greeks, but this in no way lessens Newton's accomplishment. He was the first to *prove* the force law, by mathematical derivation and comparison with observation. Furthermore, he followed up this proof with detailed applications to a large variety of celestial and terrestrial phenomena. No one else did this, and perhaps no one else could have.

The geometric analogy does not guarantee that force laws must be of an inverse square type, but it illustrates why people found it to be a natural kind of distance dependence. It also suggests that there is a close link between gravity and geometry. The true nature of this connection was not understood until more than two centuries later when Einstein worked out the general theory of relativity.

There is a remarkable property of the inverse square law that bears special comment. *If gravitation falls off inversely with distance with any power dependence other than two, stable orbits are not possible.* That is, if the force of attraction goes as the reciprocal of distance to the power of 1.8 or 1.99 or 2.001, or 2.2, or anything other than exactly 2, the planets cannot establish any stable orbit around the Sun. If the power is greater than 2, the planets would crash into the Sun; if the power is less than two, they would fly off into space.

Halley's frustration at the inability of the coffee house trio to prove the inverse square law led him to prevail on the only man alive that could do it. Only Newton realized that a new and rigorous mechanics was needed. The science he created held mysteries and wonders that were just as great as those of gravitation. It is embodied in Newton's three laws, which are simple statements about the response of physical objects to forces. These are:

1. The law of inertia:
Every object at rest, or moving with constant velocity, persists in its state of rest or uniform motion unless acted on by a force.

2. The force law:
The force acting on a body of constant mass is the product of its mass and its acceleration.

3. The law of action and reaction:
If two bodies interact, they exert equal and opposite forces on each other.

Since Newton's time, the description of matter based on these laws has become so common and so deeply entrenched in our consciousness that it is now regarded as intuitively obvious. Indeed, the surprising results of modern theories are often presented in counterpoint to those of classical Newtonian mechanics, which are taken to be just common-sense science. But at the time, Newton's mechanics was even more revolutionary than either quantum theory or relativity, which, in a very real sense, actually validate Newton except for atomic-sized particles, the universe as a whole, and for speeds close to the velocity of light.

Even the language to properly describe motion did not exist before Newton. He had to refine the concepts of space and time, of force, mass, and momentum, of inertia, and of the addition of multiple forces before he could describe motion to the degree of rigor demanded by his high standards. Only after he had set the mechanics of motion on a secure base would he accept his proofs of the law of universal gravitation.

The consequences of Newton's mechanics are so rich and far reaching that it was the subject of expansion, analysis, application, and verification for the next two hundred years. It is still the foundation of much of modern science and technology, from mechanical engineering to space exploration.

6

The laws

The law of gravitation is awesome because it controls all the matter on Earth and organizes the entire universe. But it is not a solitary wonder. For its operation in the world, it depends on the ways in which material bodies can move and on the forces that can act on them. It was part of Newton's genius to realize that an understanding of forces and motion was essential before the inverse square law could be demonstrated or applied. His work gave us the three laws of motion that bear his name, and, while not as dramatic as the law of gravitation, each is as universal, and as much of a mystery.

The first law is deceivingly simple. It states that, if an object is sitting still, it will not move unless a force acts on it. A stationary ball on a billiard table will not move unless it is hit by the cue stick, or by some other ball. This is so obvious it barely needs comment. Of course it will not move unless it is pushed. Why should it? And of course it will move if a force is applied. But there is another part to the idea of inertia, which states that if a body is moving at a constant velocity, it will continue to move at that velocity unless a force acts on it. This is less obvious. When we watch a ball moving on a billiard table, it does not go on forever, and there is nothing on Earth that moves forever in the absence of forces. The modern answer to this is that if the retarding force of the cloth on the ball did not exist, the ball would indeed continue to move.

The truth of this statement is certainly not self-evident. In fact, it is contrary to our observations in everyday life. We never see anything moving that does not slow down. Even a hockey puck on smooth ice slows and stops sooner or later.

We see many stationary objects, and to get them moving, some force must be exerted. These observations were the basis for Aristotle's ideas on motion. He believed that the law of inertia would hold true in completely empty space, but he did not conclude this by abstraction to perfectly smooth surfaces. He merely said that in a vacuum, there would be no reason for a body to change its state of motion. But since he knew from looking at the real world that a continual force was needed to keep an object moving, he concluded that there could be no vacuum, since in a vacuum, objects would move without the application of a force. Ordinarily, he said, a body can continue moving only if it is propelled by something else. In a void, however, there is no "something else" and there is no reason why a stationary body would ever move. Similarly, there is no reason that a body in motion would ever stop. Thus, a body at rest would stay at rest and a body in motion would stay in motion. (Note that this only requires the insertion of the term "uniform motion" to give Galilean–Newtonian inertia.) If this were true, then bodies in a vacuum could move without any cause. Since, for Aristotle, this was impossible because everything had to have a cause, he concluded that there could be no vacuum. Aristotle's mechanism for the cause of motion seems illogically contrived to the modern ear. He assumed that when an object is moving through air, somehow the air rushes around to get behind it and push it forward. It makes us feel as if he would do anything to satisfy a philosophical principle.

In this argument, Aristotle held that *any* motion requires a continual force acting on the body, not just accelerated motion. Aristotle found this compelling, and it led to his famous dictum that "nature abhors a vacuum". Why and how things moved was the subject of much Greek thought, because they rightly believed that understanding motion was a prerequisite to understanding the physical world. Not all thinkers agreed with Aristotle. The atomists held a contrary view, maintaining that all phenomena were the result of different arrangements of different kinds of atoms moving in empty space: hence their proposition that "all is atoms and a void".

Aristotle's ideas won out. He wrote and taught so extensively on every subject from biology to cosmology, knew so much, was so logical, and was so insistent that observation of nature was critical, that he was hard to deny. Also, Thomas Aquinas managed to integrate Aristotle's physics and cosmology into Christian doctrine, so his acceptance in post-medieval Europe was guaranteed.

Let's return to Galileo's analysis of motion on a ship, remembering that the ship is presumed to be sailing on such a smooth sea with such a uniform velocity, that if we close our eyes we could not tell we were moving. Many of us have had such smooth airplane flights that this is easy to imagine. Looking at a stationary object on the ship, we know that there is no force acting on it, but we also know that relative to the shore, it is moving with a constant velocity. We conclude that there is no force acting on an object whether it is stationary or in uniform motion.

Here is a trivial clarification: velocity has both magnitude and direction, so a body moving with a constant velocity is moving along a straight line at a constant speed. A force is needed to change either its speed or its direction.

The law of inertia is quite simple, but it is essential for understanding the orbits of the Moon and the planets under the influence of gravitational forces and for Newton's demonstration that gravity acting on terrestrial objects and on the Moon is identical. It is simple, but profound.

The motion of a body in a gravitational field is illustrated by a thought experiment just like Newton's for getting the gravitational attraction of the Moon. It is worth repeating. Imagine standing on a tall tower on the Earth and throwing a ball straight out from it. Also, pretend there is no atmosphere so there is no retarding force from the air on the ball. The ball starts moving straight out because of inertia, but the gravitational force accelerates the ball downward, so its path does not remain linear. The combination of the constant inertial motion and the gravitational accelerated motion results in a parabola as the ball rushes to the ground. The distance the ball travels before it lands depends on the height of the tower and on the initial velocity of the ball. The higher the tower and the harder the ball is thrown, the longer it will take to fall and the further will be the distance from the tower. It is easy to see that, if the tower is high enough and the ball is thrown hard enough, it will fall so far from the tower that it will miss the ground entirely. The resulting orbit is the result of the gravitational force, directed towards the center of the Earth, and the linear inertial motion in a direction perpendicular to it.

This argument is readily extended to the motion of the Moon around the Earth or the motion of the planets around the Sun.

That's all there is to it! If we accept linear inertia and universal gravitation, we can calculate the orbits of celestial objects.

Let's go back again to Galileo's ship on which a ball dropped from the top of the mast. To sailors on the ship it will look like it is falling straight down, although its path looks like a parabola to someone watching from the shore. Furthermore, everything else that happens looks normal to the sailors. The galley fire, the coiling of ropes, the sailor's walk: all look just as they would if we were on shore. Our modern experience in planes is similar. If we drop something, it falls straight down, our coffee pours just as it does on land, and a walk in the aisle is just as it would be in the airport.

All the laws of physics are the same on the ship as on the land, provided only that we measure everything relative to the ship and not to the shore. This is a more general statement of the law of inertia than Newton's first law, but includes it if we grant that all the laws of physics are ultimately based on the analysis of motion. These simple observations are the seeds of those profound revisions of our ideas of space and time embodied in the theories of special and general relativity.

The extension of the law of inertia can be stated precisely and in its full generality only by using the idea of coordinate systems as described above for analytic geometry.[27] It is a simple idea. Consider the measurement of a length of board needed, say, to construct a shelf. The correct procedure is to take a ruler, lay it along the board such that it starts at one end, measure off the needed distance, say one meter, and make a mark on the board where it is to be cut. The distance is measured relative to the starting point. If the ruler were placed such that the board's edge coincided with the half-meter mark on the ruler, then we would mark off the end to be cut at one and a half meters. The length of the cut board is still one meter. Relative to the first point, the ruler gave a measure of one meter, while relative to the second point it gave a measure of a meter and a half. That is, to make measurements of distance, we must have a point of origin. It doesn't matter which point in our workshop we take as origin, since distance is always a difference between two points. The only requirement is that whatever point we choose, it must be stationary with respect to the workshop. We would not get a valid distance if the point of origin were on a rolling wheeled table while we were standing still.

A familiar example of a coordinate system in two dimensions is a sheet of ordinary graph paper. The two-dimensional grid is used to identify points and measure distance. Place a pencil dot at any point on the grid. If we take the lower left-hand corner of the grid as origin, then the location of the dot can be specified by two numbers. Assume that you placed the dot at the intersection of the second vertical line with the fourth horizontal line, then the location can be written as (2, 4) because it takes two steps

[27] The word "system" is used in a number of different ways in science, the precise meaning being clear from the context. In thermodynamics or statistical mechanics, for example, a system can be any material object. In astronomy, the Sun and the planets, along with their moons, are called the solar system. Anything that consists of a number of related parts is called a system. A coordinate system is simply the aggregate of points in a line, in a plane, or in space.

to the right and four steps up to reach the point from the origin. The distance from the origin to the point is obtained from the Pythagorean theorem.[28] Just square the two numbers, 2 and 4, add the squares, and then take the square root of the sum. Then we find that the point is approximately 4.5 units from the origin. The two-dimensional graph paper grid is a two-dimensional coordinate system, and every point on it is located by two numbers that are its coordinates.

A three-dimensional coordinate system is similar. Once a point of origin is chosen, the distance to any other point can be described by the three ordinary coordinates of analytic solid geometry which form a three-dimensional grid that gives a method to quantitatively describe where things are. To locate an overhead light fixture, for example, choose a lower corner of the room. Starting from that corner, walk five feet along the wall, make a perpendicular turn, walk another seven feet, and stick a rigid ruler straight up into the air and note that the fixture is six feet up. The coordinates of the fixture are (5, 7, 6) and give its location in the room. Note that choosing a different origin would give a different set of coordinates. However, the distance between the light fixture and a chair would be the same no matter what we chose for the origin. Distance is invariant with respect to the changes in origin. There is a possible coordinate system for every point of origin in the workshop, and they all give the same answer for a measurement of the distance between two objects, provided they are all stationary with respect to the walls of the workshop and therefore stationary with respect to each other.

A coordinate system is just the set of all points that can be measured from a fixed origin. It is also often called a frame of reference or simply a reference system

[28] I state the Pythagorean theorem here so the reader need not go back and look at a plane-geometry textbook. It refers to a right triangle, which is just a triangle one of whose angles is ninety degrees. Each side that ends at the right angle is simply called a "side", and the side opposite the right angle is called the hypotenuse. The theorem states that the square of the hypotenuse equals the sum of the squares of the two sides.

Returning to Galileo's ship, it is clear that someone on shore would make distance measurements relative to a stationary Earth while someone on the ship would make measurements in a coordinate system that was stationary with respect to the ship. Each carpenter would have the good sense to measure his board in the locally stationary coordinates. Now let's restate the law of inertia as follows:

The laws of physics in reference systems moving relative to each other with constant velocity are the same.

The observations were obvious, the thought experiments and generalizations straightforward, and the logic was simple. Yet the law of inertia has implications that are of such far-reaching importance that it belies its unassuming origins. It contains specific assumptions about the very nature of space and time, assumptions that we live by and have accepted as universally correct since the time of Newton.

It was Einstein's re-examination of the law of inertia that led to special relativity and the clarification of the meaning of space and time measurements in moving systems.

In modern texts, Newton's second law is usually given in terms of a continuous force acting on a body of a given mass with the statement that the force is the product of mass and acceleration. But this is not the law as Newton gave it. His second law did not refer to continuous forces, but to impulsive forces. It was obvious to Newton that the impulsive force needed to change the velocity of a body depended not only on its velocity, but also on how much matter the body contained. If a certain force was needed to change the velocity of a chunk of lead by a certain amount, then surely it must take double that force to bring about the same change for a chunk of lead twice the size.

The idea that both velocity and the amount of moving matter were needed to describe motion is quite old, and this evolved into the idea of mass. There must be some property that describes the amount of matter in a way that is independent of composition. That is, some measure of that amount must exist that applies equally to lead, clay, salt, water, or any other material. The idea of weight was well established, and any material could be cut into one-pound chunks, but, because weight varied from one place to another, it could not be the ultimate definition of the amount of matter. The definition of mass was vague even after Newton's work, but its importance was well recognized. The product of mass and velocity, which we know as momentum, was called "the quantity of motion" and was at the center of the development of dynamics. Descartes's, on theological grounds, held that the quantity of motion had never changed since the creation of the world and tried to formulate a principle of conservation of momentum. Almost everything in Descartes physics was wrong, and this included his momentum conservation rule, which did not recognize that velocities after a collision had opposite signs to those before the collision. Nevertheless, the conservation of momentum was taken to be a fundamental law of nature on which to base the science of moving objects. Leibniz and his followers opposed this view. They believed that the important quantity was not the product of mass and velocity, but the product of mass and the *square* of the velocity. This was called *vis viva* (living force), which we recognize as proportional to the kinetic energy. Huygens made the most careful analysis of colliding bodies and found that *both* momentum and kinetic energy were conserved for elastic collisions, so the battle between the two schools was meaningless.

It was natural for Newton to take momentum as the fundamental entity in describing motion, and his second law states that an impulsive force acting on a body equals the change in momentum resulting from the impulse. This is more general than the form taught in elementary physics courses. (Force equals mass times acceleration.)

Newton was fully aware of the continuous form of the second law, in which the force is the rate of change of momentum, and in fact derived it from his statement in terms of impulse forces by a limiting process in which sequential impulses are applied to a body over ever decreasing time intervals.

Many of Newton's predecessors, going as far back as the ancient Greeks, knew that velocity alone was not enough to describe moving bodies and introduced ideas analogous to momentum, although there was no clear definition of mass. Historical analysis suggests that one reason Newton wrote the second law first in terms of impulse was because impulsive forces, as arising from collisions, were well understood and accepted, whereas continual forces could be attributed to celestial bodies only by adopting the idea of action-at-a-distance. Most scientists of the time could not accept this, and Newton himself was greatly troubled by the idea of gravitational forces acting over great distances without any mechanical way of transmitting them. For the *Principia,* his solution was to ignore the problem, or rather to set it aside for future consideration, by making his famous statement that he did not make hypotheses. By this he meant that he would not inquire into the ultimate causes of gravity; he just accepted it and worked out its consequences. But he did use impulse instead of continuous force in the second law, showing how it could be applied to continuous forces later in the text.

The third law is simply stated. If two bodies interact, they exert equal and opposite forces on each other. The importance of this law can hardly be overstated. If the Earth pulls the Moon towards itself, then the Moon pulls the Earth with a force of equal magnitude but in the direction towards the Moon. Celestial mechanics could not exist without this law, and it is indeed essential for all of dynamics. It has, in fact, been called the most important of Newton's laws and is the only one that Newton introduced without any precedents to help him. Newton himself attributed his first two laws to Galileo, which is not quite historically accurate. Galileo had a fine appreciation of inertia, and his idea was adopted by many. But while Galileo's study of motion and falling bodies certainly related force and change of velocity, he did not generalize this to a universal law of motion. Nevertheless, Newton was right in the sense that Galileo's work led directly to the first and second laws.

A most remarkable aspect of the laws of motion is that they introduced the concepts of mass and force without any rigorous definitions of what they really were or how they could be measured. Such rigor was to come much later. Yet Newton applied these laws to important mechanical phenomena, exactly as if the fundamentals were already established. And he applied them correctly! His ideas on force and mass were the result of a lot of analysis and intuitive insight of what force and mass could mean. The lack of rigor of definition was amply compensated by the detailed rigor of the proofs of their consequences. Mechanics was not the only case in which successful application outstripped proper analysis of foundations. The calculus was used to great advantage for many years before mathematicians could put it on a sound formal basis.

The greatness of Newton's genius lay in recognizing that *all* of mechanics could be constructed on nothing more than his three laws. He took them as axioms valid for both celestial and terrestrial phenomena, and then showed how to apply them to such an enormous variety of physical cases that they amounted to a complete solution of the outstanding problems of the mechanics of his time.

Newton captured the essence of gravity in a most important sense. He showed how all known gravitational effects could be calculated from a single, unifying theory. But his theories led to mysteries that were even more puzzling than the law of gravitation itself. Until these were dealt with, it could not be said that gravitation was understood.

The origin of the inverse square law of gravitational attraction was completely unknown. The law was deduced from observation, particularly Kepler's laws of planetary motion. It had no sound conceptual foundation. Newton's work left unanswered the question of where gravity came from and why the force law was an inverse square. And, given that gravity exists, how is the force transmitted over enormous interplanetary and interstellar distances from one body to another, with nothing between them? At an even deeper level, Newton's mechanics described the actions and movements of objects in time and through space. Newton himself said that time and space were so well understood that they did not really need definition, at least not in the same sense as mass and momentum. Yet, he must have felt a need for additional rigor because he spent some effort in distinguishing "true" time from that which is measured on clocks and "absolute" space from that which is measured relative to material objects. In the end, he succeeded only in carefully articulating the generally accepted intuitions of his time. Until the reasons for the very existence of gravitation, its action through empty space, and the precise meaning of space and time were clarified, the mystery of gravity would remain. Newton had discovered how, but there was still a great, unknown why.[29]

It is time to state the consequences of inertia more precisely as follows: the laws of physics are the same for all coordinate systems moving with constant velocities relative to each other. This is so important that some elaboration is desirable, even if it is repetitive. We want to be precise so we start with the definition: an inertial coordinate system is one in which all objects are stationary, or are moving with constant velocity, unless they are acted on by a force. Note that any coordinate system moving at constant velocity relative to any inertial system is also inertial. The reason for this is that if any object whose velocity is found to be constant (or zero) when measured in the first system, it is found to have a constant (or zero) velocity when measured with rulers attached to the second system. The difference between the two measurements is just the relative velocity of the two systems.

Consider another thought experiment just to make sure of the meaning of this. Let two trains on adjacent tracks be moving with different speeds. (Of course, these are idealized trains and tracks, with no friction, bumps, lurches, or jogs, so they give perfectly smooth rides) The people in both trains have all the tools they need to make measurements and the people in one train make all measurements relative to a train wall, which they assume is

[29] There is *always* an unknown "why"?

stationary. In the same way, the people in the other train make all measurements relative to their train wall, which *they* see as stationary. In slightly more mathematical language, the observers in each train make measurements in coordinate systems attached to their train. For each set of observers, their train is an inertial system, and the basic laws of physics are the same in each system.

There are obviously an infinite number of inertial systems, and physical laws are the same for all of them. For classical mechanics, this means that Newton's three laws hold true for all inertial systems.

It is often necessary to deal with systems in motion. Trains and ships and planets and stars, as well as electrons in accelerators and picture tubes and semiconductors, are all moving with various velocities, so we must know how to calculate the time and position of an object moving relative to us. That is, we need a method of transforming the time and space coordinates that we measure in our coordinate system into the time and space coordinates when measured in the moving coordinate system. In Newtonian mechanics, this is very easy. The time is just taken to be the same in both systems, and the position is related to the velocity between the systems. That is, if we look at a given point that looks stationary to us, we simply measure its distance from the origin of our system. Someone looking from the moving system sees that the point is not stationary but moving with a velocity that is equal and opposite to that of the second system. The observer in the second system finds it easy to calculate the position of the point as a function of time. The relative velocity of the two systems therefore defines a transformation between positions measured in the two systems. The transformation for time is even easier. The time measured in the second system is taken to be identical to that measured in the first.

These rules for connecting time and distance measured in moving coordinate systems are called the Galilean transformation, because they follow directly from Galileo's principle of inertia.

Newton's laws of motion are enormously powerful. They allow us to understand, predict, and apply natural law to a great range of circumstances. The motion of aircraft, automobiles, trains, rocket ships, and satellites, and the workings of all the mechanical parts in them; the structures of buildings and bridges; the operations of tools from screwdrivers to snow-blowers and lathes; golf clubs, baseballs, artillery shells: all of these, as well as the motion of celestial objects, from the Moon to the stars, are subject to the laws of motion. How remarkable that so much of the ever moving, ever changing aspects of a dynamic universe can be described by so few and such simple statements!

One of the great achievements of Newtonian mechanics is that it has identified the few things that are constant throughout the shifting interactions among material objects. These are called conservation laws because they remain unchanged no matter what the changes in velocity or position. They are: the conservation of energy, the conservation of linear momentum, and the conservation of angular momentum.

Energy is just the ability to do work; linear momentum of a material object is just the product of its mass and velocity; angular momentum refers to a rotating object and is related to the product of its mass and the velocity with which it is rotating. For a single mass rotating around some point, the angular momentum is just the product of its ordinary momentum and its distance from that point. No matter how a collection of objects moves around, collides, and bounces, these three quantities stay the same.

The conservation laws have both a conceptual and a practical importance. Nature, even in its simplest physical manifestations, displays continual motion, and Newton's laws can describe that change. At the same time, it is important to know if *anything* stays the same while all this change is going on. The answer is yes! There are some attributes that are constant, some properties that are always the same and therefore must be saying something fundamental and significant about nature. These are the conservation laws.

Their practical utility is apparent when applied to actual physical situations. The conservation laws are functions of masses, forces, position and velocity. Their constancy makes finding the answers to specific questions easier. For example, when two balls collide, their motion after the collision is easy to get because their total momentum must be the same before and after the collision.

The constancy of energy, of linear momentum, and of angular momentum flows directly from Newton's laws, but the conservation of mass does not. It is merely assumed. When Newtonian mechanics is modified by special relativity, it turns out that the mass is not a constant, but varies with a body's velocity. The laws of conservation of mass and of energy are then modified to a law for the conservation of the sum of the mass and the energy. But this modification is needed only for very high velocities. For everyday life and everyday astronomy, Newton's laws work just fine.

7

The system of the world

The planets and their relation to the Sun were a mystery from ancient times that demanded an explanation. Such a grand, breathtaking thing required an equally imposing name. The Sun, the planets and their moons, and the comets collectively were known as "The System of the World". The title meant more than just the collection of celestial objects themselves; it included the models and theoretical constructs devised to make sense of their motions and interrelations.

Galileo split the history of both science and religion in two by his dialogue on the systems of the world, and Newton finally gave the correct description in the third book of his *Principia*.

Modern astronomy began when Newton showed that Kepler's laws of planetary motion followed from his mechanics and the theory of universal gravitation. Let's recall Kepler's first law:

1. The planets move in elliptical orbits with the Sun at one focus of the ellipse.

The first law seems almost trivial in that it merely states that the planets all go around the Sun in geometrically similar paths. Yet it was a radically new result. The Greek idea of the circle being the most perfect geometric figure had been absorbed into astronomy over the centuries, and planetary orbits were assumed to be perfect circles right up to Kepler's time. The fact that they are *not* circular was another realization that the new astronomy and physics was quite different from the Ptolemaic–Aristotelian–Christian corpus that had been accepted for so long.

Let's take a closer look at these elliptical orbits. An ellipse is just a circle that has been squashed down to resemble the cross section of a football, or a symmetrical egg. An accurate picture of an ellipse can be obtained by the following simple construction: tape a piece of paper onto a board and stick two pins into it, and then tie a loose loop of string from one pin to the other. The string needs to be long enough to be rather slack. Now take a sharp pencil, put its point against the string and pull it out until the string is held in a tight triangle by the two pins and the pencil point. With the point touching the paper, sweep the pencil around the paper while keeping the string taut. The pencil will trace out a closed curve on the paper. The curve is an ellipse, and indeed it looks like a squashed circle. If the pins are close together, the curve will be very much like a circle, whereas when they are far apart, the curve will be long and thin. The two pins are the foci of the ellipse. When the two foci are at the same place, the curve is a circle, and the further apart they are, the greater the deviation from circularity. The amount of this deviation is called the eccentricity. This construction is often used to illustrate the definition of the ellipse as the locus of all points lying in a plane such that the sum of their distances from two fixed points is a constant.

Kepler found precisely these shapes of the orbits for the six known planets of his time. When Uranus, Neptune, and Pluto were discovered, their orbits were also found to be elliptical. (Furthermore, the orbits of planetary moons all describe ellipses around their planets.)

The planetary orbits have widely varying eccentricities, being almost a perfect circle for Venus and Neptune to an enormous twenty-five percent eccentricity for Pluto, whose orbit is so eccentric that its distance of closest approach to the Sun is slightly smaller than Neptune's, while its furthest distance is over sixty percent greater. Fortunately, the Earth's eccentricity is small, less than two percent, so that its relatively constant distance from the Sun precludes very large annual temperature variations. If Earth's orbit had the same eccentricity as Pluto's, during one year it would reach as far in towards the Sun as Venus and as far out as Mars, making the evolution of life impossible.

The common factor that makes all the orbits elliptical is that the planets are all attracted to the Sun according to the universal law of gravitation.

Kepler's second law is a kinetic as well as a geometric statement, about planetary motion:

2. As they move around the Sun, the planets sweep out equal areas in equal times.

That is, Kepler found that if the orbital distance a planet travels in a unit of time is measured, the area swept out by the orbit is always the same. Choose two points on the elliptical orbit and draw lines from them that connect to the Sun. The result is an area sector shaped like a wedge. It takes a certain time for the orbit to move from one of these points to another, and the sector has a certain area. No matter where the two points, if the time to get from one to another is the same, the swept-out area is always the same. This is summarized by saying that the areal speed of a planet in an elliptical orbit is a constant.

For a perfectly circular orbit, the length of arc traversed per unit time is constant; for an elliptical planetary orbit, this is not true, but the *sectoral area swept out per unit time* is a constant. Kepler found this to be true for each of the planets he knew, and it was correct for the planets discovered after his time. The validity of Kepler's second law for every planet again points to the existence of some basic, universal cause.

The third law has a character different from the others. It states that:

3. The squares of the orbital periods of the planets are proportional to the cubes of the average distances from the Sun.

A planet's period is just the time it takes for the planet to go around its orbit once, and this, as well as the distance from the Sun, was known for each planet from Tycho's data.

Kepler's first two laws referred to the orbits of each planet and, while finding regularities common to all the planets, they described the properties of each orbit individually. The third law, on the other hand, directly links *all* the planets. It states that if the length of a planet's year is squared and divided by the cube of its average distance from the Sun, a number is obtained *that is the same for all planets.* So not only does each planet's orbit show remarkable regularities, but the solar system as a whole obeys a simple rule of planetary motion.

It is hard to contemplate these three laws without concluding that something powerful and profound must be at work.

These laws are directly linked to the inverse square law of gravitational attraction. Using Newton's mechanics and the inverse square law, it is easy to show that planetary orbits must be ellipses. Actually, the mathematics shows that gravity allows the motion of a body being attracted by the Sun to be a parabola or a hyperbola, in addition to an ellipse. The energy or velocity of the mass determines which of these curves are followed. For the planets, the orbits are all ellipses. For some comets, the curves are parabolas, so the comet comes into the solar system, swings around the Sun, and leaves. Hyperbolic trajectories occur when the gravitational field of a planet is used to accelerate a space vehicle by the "slingshot" effect. The vehicle is aimed to pass near a massive body and speeds up as it gets closer. Its speed, which is the sum of its initial speed and that due to the acceleration caused by the planet's gravitational field, gets very high. Too high, in fact, to be caught into an elliptical orbit, so it speeds away as if thrown from a slingshot.

Remarkably, the inverse square law accounts for all of these.

The inverse square law also accounts for Kepler's other two laws of planetary motion.

Scientists had been searching for a proper description of "The System of the world" for centuries. Kepler and Newton gave it to them.

Newton's derivation of Kepler's laws applies to any orbiting bodies, not just the solar system's. The equations contain the

Table 7.1 The solar system

1	2	3	4	5	6	7	8
Planet	**Period Years**	**Perihelion (AU)**	**Aphelion (AU)**	**Average Distance (AU)**	**Eccentricity**	**Thrid Law Ratio**	**Percentage Deviation**
Mercury	0.241	0.31	0.47	0.39	0.205	0.979	2.087
Venus	0.615	0.718	0.728	0.723	0.007	1.001	−0.077
Earth	1	0.98	1.02	1	0.020	1.000	0.000
Mars	1.88	1.38	1.67	1.525	0.095	0.997	0.343
Jupiter	11.9	4.95	5.45	5.2	0.048	1.007	−0.713
Saturn	29.1	9.02	10	9.51	0.052	0.985	1.543
Uranus	84	18.3	20.1	19.2	0.047	0.997	0.309
Neptune	165	30	30.3	30.15	0.005	0.993	0.664
Pluto	248	29.7	49.3	39.5	0.248	0.998	0.204

masses and correctly describe, for example, the motion of the Moon around the Earth.

Table 1 displays some data that give the main features of the planets. The first column identifies the planet and the second column gives its period in years. The third and fourth columns give the planet's closest and furthest distance of approach to the Sun during its period of revolution. The fifth column gives the average distance from the planet to the Sun, and the sixth column gives the eccentricity. The seventh column gives the Kepler third-law constant, which is obtained by dividing the square of the period by the cube of the average distance.

The first thing to notice is that the solar system is really big.[30] Pluto is nearly four *billion* miles from the Sun, so the solar system

[30] The size of the solar system depends on what objects are defined to be in it. For example, a new object (called Eris) has been recently discovered with a highly eccentric orbit further away than, and larger than that of, Pluto. Also, a number of other objects of similar size are known to be orbiting the Sun at greater distances. Nomenclature has recently been clarified by defining a planet to have to be large enough for its gravity to make it spherical and to have cleared its surroundings of smaller bodies. For the purpose of illustration, I have adopted the simple classical definition that the solar system extends to Pluto.

is almost eight billion miles across. It takes light almost six hours to get there, so communications from Earth to a vehicle going to Pluto are seriously delayed. Light is very fast. NASA's probe, *New Horizons*, will take nearly ten years to get to Pluto, even though at nearly 30,000 miles per hour it is the fastest space vehicle ever.[31]

Next, notice that the eccentricities are small. Most of the planets have an eccentricity close to zero, which means that their orbits are close to being circular. Mercury and Pluto are the dramatic exceptions.

The seventh column is the most interesting because it verifies Kepler's third law. The deviations from Kepler's third law are shown in the last column. Of course, there must be deviations, because, as shown by Newton, Kepler's third law is strictly true only when a planet is revolving around a mass so large that its own mass can be neglected. The law of gravitation states that not only is the Sun pulling on the planet, but the planet is also pulling on the Sun. The net result is that the Sun and the planet both go around a point that is not precisely at the Sun's center. Also, all the planets exert gravitational attractions on each other, and the derivation of Kepler's laws ignores these effects. But they can be calculated, and when they are, it is found that the results are in accord with observation. This is a double triumph for Newton's laws and the theory of gravitation. Not only do they reproduce Kepler's laws, they even correctly account for deviations from them.

Of course, motions of the moons around the planets, the paths of space vehicles, and the motion of nearby stars relative to each other are all correctly given by the law of gravitation.

[31] The distances in Table 1 are given in Astronomical Units (AU). One AU is just about 93 million miles, which is the radius of the Earth's orbit. The perihelion is the closest distance of approach of the orbit of the Sun, while the aphelion is the furthest.

8

Force and mass

Newton knew that a rational theory of mechanics had to start with concepts of force and mass as well as of space and time. He had rigorous definitions of none of these four quantities and yet was able to construct a theory of mechanics that accounted for the planetary motions as well as for all known mechanical phenomena on Earth. His remarkable physical intuition led him to the correct results in spite of later research showing that space and time were *not* absolute and in spite of the lack of sound foundations for the concepts of force and mass.

The first quantity we need to clarify is Newton's "quantity of motion", which we will henceforth refer to by its modern name of momentum, and restate the second law in modern language by saying that the force acting on a body equals the rate of change of its momentum. Momentum is just the product of mass and velocity, so we need to have a clear notion of the meaning of mass. One approach is to use the second law itself to define mass, by asserting that it is just the proportionality factor between force and acceleration. Thus, if a body is accelerated, we can get its mass by measuring the acceleration and dividing it into the force acting on the body. But this is a circular definition, because it has no rigorous definition of the concept of force. If force is well defined, then the second law can be used to define mass, and if mass is well defined, then it can be used to define force. But at least *one* of the two must be defined independently of the second law.

Newton tried to get at this by working on the mass. The idea of mass, as a measure of quantity of matter, was an old one. It seemed clear that there was such a thing as "quantity of matter" which did not depend on the nature of the material bodies, and some measure of this was needed. It also seemed clear that weight was not a satisfactory definition for this mass, particularly after it was found that the same object had different weights at different places, such as in a valley or on a mountain. As in other instances when a basic definition was needed but not forthcoming, Newton talked around it. At the time, the density of objects was thought to be an important basic property of objects, primary in its own right, and not derived from its modern definition of mass per unit volume; and Newton tried to clarify the idea of mass by considering density and volume together. Of course, this was another case of circular reasoning. [32]

The solution was ultimately found by the following logic. Mass is a quantity in physical statements, so its definition must start with physical experiments. It is intimately tied to the idea of force, so let's look at bodies being moved by forces. But, to avoid circular reasoning, the definition of mass cannot involve the definition of force. The purely kinematic notions of position, velocity, and acceleration are well known and rigorous, and they do not depend on any concept of force, because they involve only position and time. So let's do an experiment in which we have two bodies, called A and B, that act on each other through some force. We do not need to know anything about the force, except that we know of its existence because the bodies are accelerating. It might be a gravitational force or a spring; it doesn't matter because we are only going to measure accelerations. Unless the two bodies are totally identical, we will find that their accelerations are different. This gives us our definition of mass. *The ratio of the masses of the two bodies is defined to be the inverse ratio of their accelerations.* Experimentally, we find that for two given objects, the ratio is a constant, no matter what kind of forces are acting. Whether the

[32] This might have been a leftover from the early Greek theory of the four elements that constitute matter. In this theory, density was of crucial importance.

experiment on the two bodies is done with the force of gravity or of springs, electric charges, or magnetic poles, the ratio is the same for any kinds of forces and any values of the accelerations. This ratio must therefore say something fundamental about the response of material bodies to forces. A numerical measure of mass is readily constructed if some standard body is taken to have unit mass. The value of the mass of any other body is then the ratio of the acceleration of that body with respect to that of the standard unit body.

This definition of mass arose from the efforts in the eighteenth and nineteenth centuries to rid Newton's mechanics of all ambiguities and put it on a logically coherent axiomatic basis. It was first clearly stated by Ernst Mach (1838–1916) in 1867 and was given a definitive form in his 1883 treatise *Science of Mechanics*. Mach was a Czech-born Austrian who is best remembered for the "Mach number" arising from his work on sound waves, in which he correctly formulated the effects of supersonic velocities. His philosophical efforts centered on his insistence that *only* experiment could be used to define physical concepts. He therefore opposed the concepts of the atomic theory of matter since atoms (at that time) could not be observed.

From its definition, we see that the greater the mass, the less its acceleration in response to a force, so it takes a larger force to move a larger mass. It therefore makes sense to multiply the mass and the velocity to get a "quantity of motion", which we call momentum.

Note that nothing about forces enters into this definition of mass except the fact of their existence, and even this is no more than a statement that the accelerations exist. Now force can be rigorously defined from Newton's second law as the rate of change of momentum. There is an important physical content in the second law that goes beyond the definitions needed for logical consistency. Physically, the second law states that if you see an object being accelerated, look for a reason. Something physical is responsible for the acceleration, and we call that something a force.

This is not the whole story, because the second law of motion is a differential equation whose utility depends on knowing the force as a function of position and of time. The equation cannot be solved without knowing this function. For example, the orbits of the planets are obtained from Newton's second law because we know that the gravitational force between two bodies varies as the inverse square of the distance. It is not enough just to have the rigorous definitions of mass and force. To use this definition as is, measurements of the acceleration between two gravitating bodies would have to be made for all the possible distances between them. But this is precisely what we are trying to find. Making all possible measurements is not possible and would only give us a table of numbers, not a theory that helps us understand the orbits. We need a functional relationship for the force. How can we get such a relation? The practical answer is that we try one out and see if it works. For gravitation, for example, we try out the inverse square law and find that the *same* law correctly describes the orbits of all the planets, the comets, and the Moon, as well as artificial satellites. Also, we find that the same law works for interacting terrestrial bodies. Then we can state that the force law is correct for gravity. A similar process works for other kinds of forces. A spring is stretched when a weight is hung from its end or when it is pulled and, since it takes more weight, or more effort, to stretch it a longer distance, we *assume* that the force is proportional to the amount the spring is stretched from its original length. This rule was in fact proposed during Newton's time and is known as Hooke's law. Using this rule to define the force as a function of distance in the second law gives equations for the oscillations of masses on springs that are internally consistent and in agreement with a large number of experiments. Therefore, we add Hooke's law to a list of forces with a known position variation. The method is general. Is a frictional force, or the resistance to a body moving in a fluid, proportional to the velocity? We try it and find out.

Note that this is *not* a tautology because when the force law is found for a particular set of circumstances, it is generalized to include a great many cases, thereby making it very useful.

An important point that is often neglected must be stressed here. The second law is often described by saying that the force is the mass times the acceleration. This formulation is encouraged by the Mach definition of mass in terms of relative accelerations and is surely true if the mass is always constant. In some cases, however, the mass is not a constant. The theory of special relativity, for example, shows that the mass of an object depends on its velocity.[33] The law as enunciated by Newton did not say that force was proportional to acceleration, but to *the rate of change of the quantity of motion*, thereby recognizing that the response of an object to a force was a change in its momentum. The basic quantity combined the ideas of mass and velocity. It is more accurate, and truer to the spirit of the second law, to define force as the rate of change of momentum, without separating out the mass.

A careful look at the role of mass in mechanics shows that it is used in three different ways. The first of these is the mass defining the resistance of a body to changes in its state of motion and is called the inertial mass. Mass also appears in the force law of gravitation, and it appears twice, once for each of two interacting bodies. The attractive force of a massive body is the result of its attracting mass, and the response of a body to a gravitational force is defined by another mass. Thus, there are three distinct masses: inertial, gravitational attraction, and gravitational response. The distinction between attracting and attracted gravitational mass is normally ignored because they both appear in the same law of gravitation, where their definitions are identical. But it is obvious that gravitational and inertial masses are two quite distinct concepts, arising from two different kinds of experiments.

It is a remarkable experimental fact that these two masses are identical. This equality became one of the foundations of general relativity.

[33] There are also important non-relativistic examples in which the mass is not constant. A rocket loses mass as it burns its fuel and a hailstone can lose mass as it falls through the atmosphere. In such cases it is essential to use Newton's law as a statement of change in momentum.

No one in Newton's time could foresee that the equivalence of gravitational and inertial mass would be a major factor in a thorough overhaul of mechanics. The existence of mass in Newton's law led to a much deeper understanding of gravity and its relationship to the structure of space.

9

Two more giants

During the century following publication of the *Principia*, mechanics flowered into an all-embracing mathematical theory, which was applied to a host of celestial and terrestrial problems with ever increasing success, widely expanding our knowledge of the physical world. However, nothing fundamentally new was added to our understanding of the laws of mechanics beyond that given by Newton. The next great wave of physical discovery occurred in a seemingly very different branch of physics.

The two nineteenth century giants of electricity and magnetism, Michael Faraday and James Clerk Maxwell, were born forty years apart and never had the opportunity to work together. They met after Maxwell was appointed Professor of Natural Philosophy at King's College, London in 1860 at the age of twenty-nine, but did not have any serious scientific interaction because Faraday was old and infirm by that time. Maxwell, of course, knew all about the older man's great work and admired him immensely. It was Faraday's thoughts on the nature of electromagnetic interactions that led Maxwell to his famous equations. They remain one of the greatest intellectual accomplishments in history and provide the base for those striking technological advances that have completely changed the human condition.

Electric and magnetic effects were known for centuries before the time of Newton; the ancient Greeks were quite familiar with the magnetic attraction of the lodestone and the action of rubbed amber or glass on bits of light material, but knew little more.

The modern study of electricity and magnetism started in 1600 with William Gilbert's publication of his *De Magnete*. Gilbert was a Fellow at Cambridge and well connected to the establishment, as shown by the fact that he was the personal physician to Queen Elizabeth I. *De Magnete* was the result of a long interest in magnetism. It summarized everything that was known, debunked a number of myths, and reported his own work. This was the first description of experiments that were more than mere observation.

Gilbert showed that a magnetic needle aligns itself in a particular direction because the Earth itself is a giant magnet, that electrical effects can be induced by friction in many other substances than amber, and that there were major differences between electric and magnetic phenomena. In keeping with the current notion that there could be no action-at-a-distance, he proposed that the action of electricity was due to an "effluvium" emanating from the electrically active material. This evolved into the electrical and magnetic one-fluid and two-fluid theories, in which electricity and magnetism were thought to be special substances.

These were great mysteries whose secrets began to be unraveled only in the middle of the eighteenth century. At the suggestion of Benjamin Franklin, Joseph Priestley performed an experiment showing that no electric force could act on a charged body inside a hollow metal ball. The analogy with gravitation was immediately obvious, since Newton's result that the mass of a hollow sphere could exert no force on any mass inside it was well known. Priestly therefore concluded that the force between charges must be like that between two gravitating masses. Charges, like masses, exerted forces that varied as the inverse square of the distance between them.

Charles Coulomb performed a series of careful experiments during the years 1785–1789 and found direct experimental proof that this was the case. He established the inverse square law for magnetic poles as well as for electrostatic charges. Like charges repel and unlike charges attract; like magnetic poles repel and unlike poles attract—all with a force that decreases as the square of the distance between them. Coulomb was able to achieve great accuracy because he had invented a torsion balance in which very small forces could be measured by noting the angle of twist of a thin fiber.

Research became more intense, and in another forty years everything was in place for a modern theory of electricity and magnetism. Volta, following up on Galvani's observation of the twitching of a dead frog's legs, had invented the electric battery, thereby making electric currents available in large quantities; the equivalence of static and current electricity was recognized; and the complex relationships between electricity and magnetism were established,

Hans Christian Oersted, in 1820, was the first to definitively show that electricity and magnetism are related. He was a distinguished Danish scientist, on the faculty of the University of Copenhagen, and found that an electric current moved a magnetized needle. Others, including Benjamin Franklin, had believed that electricity and magnetism were related, but they had not pursued their speculations. Oersted, however, followed up his observation with systematic experiments showing that the magnetized needle lined up in a direction perpendicular to the wire; the current induced a magnetic field that was perpendicular to the direction of current flow. This was the beginning of our understanding of the relation between moving charges and magnets that led to the great electrically based technological revolutions.

The discovery has been described as a serendipitous accident, but Oersted himself said that it was the result of his conviction that electricity and magnetism were intimately connected that led to his experiment.

The final experimental foundations of electromagnetic theory were provided by Faraday and by Ampere, who showed that a changing magnetic force induces an electrical voltage, and that an electric current produces a magnetic force in its vicinity. These experiments demonstrated conclusively how magnetism and electricity were closely linked.

Ampere gave the first viable theory for the action of forces produced by moving currents, thereby spurring the rapid growth of electrodynamics, the science of moving charges, which is much more complex than electrostatics or magnetostatics. Faraday later took up a series of investigations that marked him as the world's foremost experimentalist. By the time he was done, all the critically important phenomena of electrodynamics had been experimentally demonstrated, and, just as Faraday was retiring from research (about 1864), James Clerk Maxwell started on his great work, which tied magnetism and electricity into one coherent set of equations. In a few profound papers, Maxwell created and completed the modern mathematical theory of electrodynamics.

These were the two great giants of electricity and magnetism. Faraday extended and completed the necessary experimental groundwork, providing for electrodynamics what Galileo, Brahe, and Kepler gave mechanics. And Maxwell organized it all into a comprehensive theory of electromagnetism, just as Newton had done for mechanics.

Beyond the purely experimental knowledge, Faraday created a conceptual structure that was the foundation of Maxwell's theory and brought back into the center of physics the issue of how forces were transmitted through space. Michael Faraday was a phenomenon. His family was poor and the highest hopes they had for Michael were expressed by his apprenticeship to a bookbinder at the age of 14.

He ultimately came to dislike the work, and his true interests emerged early. He was fortunate, however, to be indentured to George Riebau, a decent man who took a fatherly interest in young Faraday and recognized his unusual talents. An enormous number of high-quality books in science, literature, and the arts passed through his bindery, and these were the foundation of Faraday's education. The number and variety of these books may seem large to us, but Riebau's was a substantial bindery and, in addition to new issues, it bound many individual books, because many books at that time were purchased without covers and went to Riebau's to be finished off.

Faraday educated himself by reading the books, talking with other young people of similar interests, and even performing some scientific experiments on the bindery premises. He stayed with Riebau for the seven years required to complete his apprenticeship and then took a job with another firm. He desperately wanted a scientific job but these were very scarce and he was forced to continue the career his training had prepared him for. Within a year, his fortunes changed and he entered the world of science when he acquired tickets to four lectures by Humphry Davy, the greatest chemist of that time. Private lecture series sprang up to satisfy the public interest in science, and Faraday had attended a number of them given by John Tatum, which stoked the fires of his scientific interests and brought him into contact with other scientifically minded young men. Tatum was one of a number of such lecturers, but Humphry Davy was something else. Brilliant, charismatic, handsome, and recognized as the outstanding scientist of his time, he was a star. He prepared his lectures carefully with a special attention to the spectacular, and the force of his personality held the audience's rapt attention. His public lectures were always packed. He was the outstanding figure in the Royal Institution, which was formed in 1799 to spread scientific knowledge throughout the general population.

Faraday was enthralled by the lectures. He took careful notes, bound them, and sent them to Davy with a request for a job in his laboratory. They must have been impressive because, when a chemical assistant at the Royal Institution in 1813 was fired over a fist fight with his superior, Davy hired Faraday, thereby launching one of the most remarkable careers in the history of science. Faraday's habit of taking notes started early; he had already worked up a complete set from Tatum's lectures. The notes were useful in recording and organizing what he learned. Also, Faraday was prone to memory loss. He was subject to bouts of temporary amnesia that would last for varying, usually brief, periods of time, so the notes were important.

Faraday's duties ranged from washing glassware to helping Davy with his experiments, but he still found time for work of his own and rapidly acquired an excellent reputation. He stayed with the Royal Institution all his life and succeeded to the Directorship of the laboratory upon Davy's death.

Davy and Faraday were in very different circumstances. Born into a middle-class family, Davy had risen to prominence and wealth on the strength of his abilitises and personal charm, and his marriage into nobility. He was at the top of the heap in his own and in the public's estimation and he lived accordingly. Faraday came as his assistant and, in many respects, became also his servant. Lady Davy was a pure snob and never let Faraday forget his inferior status. Yet Faraday rapidly grew into an independent scientist at the Royal institution and ultimately surpassed Davy.

At the time, the cutting edge of new science ranged over chemistry, electrochemistry, electricity, and magnetism, and Faraday contributed enormously to each of these areas. He discovered benzene and several new chlorides of carbon, established the fundamental laws of electrolysis and battery action, and studied the magnetic properties of matter, including diamagnetism. He performed the definitive experiments on electromagnetism, and this work had the most far-reaching effects. By inventing the electric generator and electric motor he started the electromagnetic technology that dominated the nineteenth and twentieth centuries. The concepts that made these inventions possible became the foundations of Maxwell's theoretical studies on electrodynamics.

Faraday was no mathematician. With no formal schooling, and with his desire to see the actual workings of nature rather than abstract creations of the human mind, his self-education did not include much mathematics. He had a highly developed sense of visualization, an ability to pay close attention to detail, and a powerful command of logical reasoning. He applied these talents to his experimental work with great success. His mathematical deficiencies did not seem to handicap him in the least. In fact, if he had been a mathematician, he might not have invented his most fruitful idea, which was the existence of lines of force.

Quite a lot of elegant mathematics had been created for electrostatics and magnetostatics. These were the sciences of stationary electric charges and stationary magnetic poles, based on Coulomb's inverse square laws. They followed a course similar to that of Newton's theory of gravitation because the force laws are mathematically identical to that for gravitation. Just as for gravitating masses, the force on two isolated charges, or two magnetic poles, varies as the inverse square of the distance between them.

The forces on moving charges are different and display the intimate connection between electricity and magnetism. A moving charge experiences a force from the electric field arising from all other electric charges, just as if it were stationary, but in addition, it creates a magnetic field, and conversely, moving magnets produce electric fields. Additional forces exist that are more complicated than a simple inverse square law. The additional forces are proportional to the strength of the magnetic field and to the velocity of the charge, and act in a direction that is perpendicular to the direction of motion.

Ampere's great achievement was to incorporate this effect into a coherent mathematical theory. Still, it was all based on the mathematical form of force laws, saying nothing about it how these forces were transmitted, so they were action-at-a-distance theories.

Gravitation was the outstanding example of mathematical theory in physics and it was based on Newton's universal law of gravitation, which described gravitational attraction as acting through empty space. Newton's ambivalence about action-at-a-distance was forgotten. He had said, "I do not make hypotheses" to stress that he was just looking at the consequences of universal gravitation and not at how it was transmitted. This attitude was carried over to electrodynamics.

Faraday thought this was not enough. He wanted a physical picture of the forces, not just their mathematical formulae, and he wanted to know what was happening *within* the space separating the charges and magnetic poles. He took his cue from iron filings. The well-known classroom demonstration, in which iron filings are sprinkled on a sheet of paper laid over a bar magnet, is a vivid picture of magnetic action. Faraday just asserted that there was something real in space that made the filings line up the way they did. Every magnetic pole, and every charge, was a source of a great set of densely packed lines of force, which emanated from the source into all of space. These were not mere abstractions. Faraday maintained that they were real and he showed how they mediated the actions of magnets, charges, and currents on each other, and how they explained all the observed electric and magnetic phenomena, from simple attraction and repulsion to electromagnetic induction. The resulting picture was detailed and involved, but it was totally self-consistent and accounted for the facts. Space was not empty. It was full of lines of force that acted on each other.

In contrast, the work of most mathematical physicists of the time said nothing about what happened in the space between charges or poles.

Faraday had created the first field theory. The important physical entity was a field of force lines in which the important action was their influence on other lines immediately adjacent to them. There is no action-at-a-distance. Let me note in passing that the mathematical representation of electromagnetism can be expressed in two ways: either by global integral equations or by local differential equations. An integral equation is just a formula that adds up all electromagnetic effects over some large volume, while a differential equation is a description of rates of change of these effects at specific points in space.

The first of these ignores the intervening spaces, while the second assumes the points in those spaces describe the phenomena. Mathematically, both representations are exactly equivalent and are easily converted into each other. Physically, however, they lead to very different pictures of nature. The differential approach, that is, lines of force, was essential for Maxwell's great synthesis. Modern mathematical physics defines a field to be just some physical quantity that is a function of position in space. Thus, we can have a temperature field, a concentration field, a stress field, and a gravitational field, as well as an electromagnetic field. In this sense it is an abstract idea, but for Faraday and Maxwell, the lines of force, and the resulting fields, were very real, and without them the correct laws of electrodynamics could not have been found.

James Clerk Maxwell was 24 years old, had just completed his course of studies at Cambridge, and had become a Fellow of Trinity College when he wrote the first of three seminal papers on electricity and magnetism in 1855, just as Faraday had completed the definitive volume of his life's work and was retiring from active scientific research. Maxwell sought the secrets of electromagnetism for ten years, publishing his third paper in 1865. These papers led to his monumental theory of electrodynamics, published in his *Treatise on Electricity and Magnetism* in 1873. A few years later he became ill and died of abdominal cancer in 1879. He was only forty-eight years old.

Maxwell's work established the concept of "field" in a way that Faraday could not. Faraday was well respected, but his mathematical shortcomings were well known and the cutting edge of theory was being expressed in ever more sophisticated mathematics. And while Maxwell adopted, and even stressed, Faraday's pictorial approach, he was an excellent mathematician whose final equations were beyond criticism.

He is justly famous for his contributions to electrodynamics, and this is regarded as his most important work. But he also made great advances in other areas of science. Early in his career, he was able to explain how the rings of Saturn could maintain their orbits over millennia. The stability of the rings had puzzled astronomers and physicists since they were first observed, and Maxwell showed that they could be stable only if they were aggregates of many small particles, an explanation that has turned out to be correct and was verified by actual observations. Another great work was on the kinetic theory of gases and the statistical mechanics that described gaseous properties. He thereby clarified the relation of molecular motion to heat and temperature, and applied statistical probabilities to the study of matter. He worked on the theory of color and was the first to show that combining three primary colors could produce color images. His interest in thermodynamics led to important thermodynamic equations that appear with his name in all modern thermodynamic texts.

The Cambridge curriculum did not include any of the modern advances in electricity and magnetism, so Maxwell set himself the task of learning the new results and ordering them into a consistent whole. He had two major sources of inspiration: the recent work of William Thomson (later Lord Kelvin) and the researches of Michael Faraday. Thomson was a multifaceted genius and a powerful mathematician who was steeped in the mathematical analysis of the Continental tradition based on action-at-a-distance. His breadth of interest led him to recognize that widely different physical phenomena were often described by equations that were of similar form. This insight revealed that a number of equations in electricity had the same form as some equations in the theory of heat. The symbols had different meanings, but the structure of the equations was exactly the same. He therefore could solve seemingly intractable electrostatic problems by looking at their counterparts in the theory of heat and pointing out that they had already been solved. Modern scientists know that a great variety of different subjects are often subject to equations that have the same structure, so that knowing one field, and learning its mathematics, often provides knowledge of several other fields. But this was not widely appreciated in Thomson's time and it was important for Maxwell's work because he learned a great deal about electrodynamics from the analogy of its mathematics to that of fluid flow.

Thomson's work certainly influenced Maxwell in two ways. The first is that Maxwell made extensive use of analogies. Just as Thomson used the known mathematics of heat conduction to solve electrostatic problems, Maxwell drew heavily on the mathematics of fluid flow and mechanical processes to create analogies with electricity and magnetism. Also, the kinetic theory said that heat was conducted by the action of atoms and molecules on one another. If the temperature varied with position, as, for example, in a metal rod heated at one end, the atoms in the hotter regions are moving faster than those in colder regions. The kinetic energy is transmitted down the rod by more energetic atoms colliding with slower neighbors, thus transmitting heat. If heat was transmitted by local transfers of energy in a material medium, and if the mathematics of electricity was like that of a fluid, why couldn't the transmission of electric force be the result of local interactions moving through some medium? A study of Faraday's work convinced Maxwell that this was the case.

Faraday's work did not get greater acclamation among theoreticians, because he presented his results in descriptive, rather than mathematical, language. But Maxwell's genius immediately saw the truth in the coherent, tightly woven interrelationships in Faraday's work, and believed that, in many ways, Faraday's thought was more "mathematical" than that of many mathematical theorists.

It was Faraday's lines of force that captured Maxwell's attention and led him away from the classical idea of action-at-a-distance. He became convinced that they were real physical things, and they immediately suggested analogies with the properties of a fluid, the lines of force being analogous to streamlines and the density of electric charge being analogous to the density of a fluid.

Let's recall the precise meaning of "action-at-a-distance" so as to appreciate the fundamental idea of fields. Certainly everyone knew that an electric charge was influenced by other, distant, charges. No one denied that charges, or magnetic poles, or gravitating objects, interacted over long distances. The key characteristic of action-at-a-distance theories is that they say nothing at all about the space between the interacting objects; only the distance between them matters. Somehow, one object instantaneously feels a force exerted by the other object. And field theories do not deny that objects that are far apart affect each other. The great difference between the two is that in field theories the forces are propagated over large distances by the successive interactions between adjacent points in space.

Maxwell's entire first paper was devoted to Faraday's lines of force and showing that they could be modeled by the flow of an ideal fluid. Although Maxwell was an accomplished mathematician, the 1855 paper was written in a purely descriptive style, with no equations or mathematical derivations, thereby following Faraday's own mode of presentation. His second paper, in 1861, showed that he regarded the lines of force as real physical entities and that mechanical models of local interactions could be constructed for electric and magnetic phenomena. In 1865, he showed how to express his physical ideas mathematically, without specific reference to mechanical models.

Maxwell's fundamental concept derived directly from Faraday, who held that purely mathematical relations were not a sufficient goal, and that a clear physical understanding underlying experimental data was required for any real knowledge about nature. This was a criticism of the modes of theoretical research prevalent at the time, which started with mathematical formulae obtained from the analysis of experimental data. The physical basis of the formulae was taken to be forces among bodies, without any consideration of the space between them, an approach that was just like Newton's. For Faraday and Maxwell, this action-at-a-distance was impossible and they looked to the intervening medium for the seat of electromagnetic action. They created the idea of a field, as a property of space, in which something exists at every point that can support a force.

Maxwell's model of the mechanical equivalent for these forces was of spinning vortices, whose centrifugal motion exerted forces on each other, thereby transferring the electrical and magnetic effects. These were similar to the vortices of Descartes, except that Maxwell got all the mathematics and physics right and therefore was able to construct a logically coherent model. The existence of vortices required the existence of a fluid to support them, and this was the ether, consisting of particles so small that they could penetrate any matter and were undetectable except through their effects. In his 1865 paper, however, Maxwell based his work on an advanced form of mechanics, so his equations were independent of any model.

Actually, much of the mathematical structure of electrodynamics was already in place. The quantitative relations between changing electric charges and moving magnets were known and electromagnetic induction was well understood. Maxwell put these into one coherent set of equations, in differential form, so they described how electricity or magnetism at a point in space was related to that at an adjacent point, and noticed that the electric and the magnetic fields appeared in an almost symmetric manner in these equations. This was expected, because a moving charge created a magnetic force, while a moving magnet created an electric potential. But there still was some asymmetry. Magnetism and electricity did not appear in exactly the same way in the equations. Using his analogy with fluid flow, Maxwell realized that the equations stated that *charge was not conserved*. If the lines of force represented a physical thing that was like a fluid, then, since the amount of matter had to be conserved, so must charge. That charge could not be created or destroyed had been proposed by Franklin and was an accepted fact, but it was not so according to the equations of electrodynamics that Maxwell had inherited from his predecessors. The solution was straightforward. He simply added a term to the equations, called the displacement current, because it arose from his understanding of Faraday's lines of force, in which electricity moved through empty space as well as through wires. An electric current, or its equivalent, existed even in empty space where there were no charges, and it was the result of a varying electric field. It made the equations symmetric and properly expressed the conservation of charge. The displacement current was only one term but it changed everything, bringing a unity and simplicity to electrodynamics comparable to that brought to mechanics by Newton's laws of motion. The mathematics captured *all* of the known electrical and magnetic phenomena and predicted new ones. Let's summarize their origin. The inverse square laws for the interaction of electric charges and of magnetic poles are responsible for all electric and magnetic properties of stationary charges or magnets. For moving charges or magnets, the forces are more complex. A moving charge creates a magnetic field *throughout space* that is perpendicular to the direction of motion. Similarly, a moving magnet creates an electric field *throughout space* that is perpendicular to the direction of the magnet's motion. This means that charges can induce magnetic forces and magnets can induce electric forces, thereby making electric generators and electric motors possible. It was the symmetry displayed by the electric–magnetic induction of forces that led Maxwell to complete the theory by introducing the crucial displacement current.

Incredibly, Maxwell's equations also showed what light was and how it was transmitted through space! The equations had a great many solutions, each corresponding to some particular set of circumstances, and one of them gave a wave equation, describing the periodic motion of electric and magnetic forces through space. That is, an equation fell right out of Maxwell's work that described the variation of the electric and magnetic fields, just as the equations for the stresses in a liquid or an elastic solid described waves of pressure or distortion. The electric and magnetic fields moved through space just as sound waves move through air, water, or metal. Faraday had thought that light was electrodynamic in nature, and Maxwell gave this idea a concrete, definitive form.

The conclusion that these waves were the fundamental stuff of light was verified by just looking at the units of measurement. A set of units had been defined, based on electric charge, to measure electric quantities, and these were called electrostatic units. At the same time, another set of units, based on the magnetic field produced by a current, was defined to measure magnetic effects. Since these were invented to describe different things, the units were quite different. But the equations of electrodynamics describe *both* electric and magnetic fields, and every wave equation contains a factor that is just the square of the velocity of the wave. In Maxwell's wave equation, this is just the ratio of the electrostatic to the electromagnetic units. These units are well known and their ratio gives a velocity for the electromagnetic wave that has exactly the same value as the velocity of light! Actually, the values of the units were not known with high accuracy at the time, so the agreement of the velocity of light with the ratio of units in the wave equation was not as close as Maxwell wished it to be. He therefore applied his formidable experimental talents to the measurement of the electric and magnetic units and found a more accurate agreement.

Maxwell's actual thought process was based on Faraday's lines of force and the extension of his concepts of the mechanism of the interaction between electric and magnetic fields, and the numbers verified his expectation. The conclusion was revolutionary. Light was nothing more than a succession of electric and magnetic fields moving through space! The process is simple. An electric wave exists that is perpendicular to the direction of motion. The varying electric field induces a varying magnetic field perpendicular to it, which in turn induces a varying electric field a little later and a little further away. This continual propagation is the electromagnetic radiation of which visible light is one example. In fact, electromagnetic waves can have any frequency or, equivalently, any wavelength.

Radio waves have low frequencies and long wavelengths. Those used in AM radio have frequencies of about a million cycles per second, corresponding to wavelengths of five hundred feet or more. FM and television use shorter waves (about ten feet long), while those in microwave ovens are only about five inches long. Radar waves are even shorter with wavelengths about one inch. Infrared waves (the primary means by which matter at ordinary temperatures radiates heat to other matter) have lengths of the order of ten thousandths of a centimeter. The visible spectrum, the portion of the electromagnetic spectrum by which we see, is very small, spanning less than an octave of frequencies, from about four to seven hundred thousandths of a centimeter. Beyond this, the ultraviolet rays are about ten to a hundred times shorter. X-rays have wavelengths about ten times the diameter of a hydrogen atom or less, and gamma rays, produced in many nuclear and fundamental particle reactions, are even smaller. The radiation we use has wavelengths that range from the length of several football fields to the size of atoms, and it is all described by Maxwell's equations.

Hertz experimentally demonstrated the existence of electromagnetic waves, and the attending displacement current, in 1888.

It would be a mistake to think that the physics of the time was an arena in which action-at-a-distance theories were in a great battle with local field theories. Nature was being confronted by men of great intellect, and most of them used either local or long-range theory as was appropriate to a specific study. Newton's second law of motion, after all, was a differential equation, local in character and describing conditions at points in space. The equivalence of local and global methods, however, was not fully appreciated, because the mathematical theorems needed to convert one to the other were not worked out until Maxwell's time.

While mathematically equivalent, the fundamental understanding of nature for the two methods is different. Most importantly, Faraday and Maxwell created a new concept of "empty" space and brought the issue of the ether once more to center stage.

Maxwell's equations were truly remarkable. It took only a few lines to write them down and they successfully described all the known properties of electricity and magnetism. At the same time, they predicted new, dramatic connections between light and electromagnetism. The predictions from the equations were quantitative, so they could be, and were, verified in numerical detail. The equations worked precisely and well and needed modifications only with the advent of quantum theory.

Only Newton's laws of motion gave such a concise core of theory from which so much of nature could be understood.

At the time, no one anticipated the critical part Maxwell's equations would have for an understanding of gravitation and the nature of space and time.

10

Ether

The Sun and the Earth are 93 million miles apart, and yet they exert powerful gravitational forces on each other. Interstellar distances are very much greater, and yet gravity acts across them. This action-at-a-distance, with nothing in the intervening space, is a hard concept to accept. So hard, in fact, that it was rejected by nearly all who thought about it. The human experience of forces is that of a muscular push or pull, or a direct impact between touching objects, not something acting through empty space. In fact, empty space, absolutely empty space, with absolutely nothing in it, seems to make no sense at all. Any reflection leads to the idea that "empty space" really means non-existence, and anything that does not exist cannot have any effect on things that do exist. There must be *something* that enables bodies to act on each other even if they are not touching. The belief in an all-pervasive substance, present everywhere and always, has ancient roots. As far back as 500 BC, Parmenides held that all things were part of such a substance.

Aristotle's conclusion that there could be no void was based on his rejection of the possibility of action-at-a-distance. Nothing could move unless it was acted upon by something else, so the all-pervasive substance, the ether, had to exist.

While not necessarily subscribing to the Aristotelian theory of motion, later investigators did believe in the ether. Both Kepler and Descartes, for example, could not accept action-at-a-distance and thought that all bodies must act on each other through the agency of pressure and contact. The action of the Moon on the tides, the forces between magnets or between the planets and the Sun, the electrical attractions or repulsions of amber or glass: all required a mechanical contact. Thus they postulated a *plenum*, a continuum filling all space (Kepler) and vortices consisting of rotating ether particles (Descartes) by which one body could press on another

In 1675, Newton wrote that "there is an aethereal medium, much of the same constitution with air, but far rarer, subtler and more strongly elastic". It is "a vibrating medium like air, only the vibrations are far more swift and minute". "It pervades the pores" of all natural bodies though having "a greater degree of rarity in those bodies than in the free aethereal spaces". In passing from the Sun and planets to the empty celestial spaces, it grows denser and denser perpetually and therefore causes gravitation.

But Newton said the ether vibrations could not constitute light because he rejected the wave theory, which would require light to "bend into the shadows". He postulated that light consisted of "small bodies" impinging on, and exciting vibrations in, the ether. These vibrations communicate heat to material objects.

The ether clarified Newton's concept of absolute space, which he thought to be the material to which an absolute reference system was attached, thereby avoiding the awkward fact that all spatial measurements were made relative to some material frame of reference and therefore were always relative.

As the wave theory gained dominance, the ether took on the role of the medium in which light waves existed. After all, where there were waves, there had to be something to support them. Just as water or air carried sound, the ether carried light, and, in that capacity, it was called the "luminiferous ether". And after Young and Fresnel concluded, from analysis of the polarization of light by crystals, that the waves were transverse, not longitudinal as previously thought, the ether was assumed to have the properties of a solid. In a transverse wave, the medium vibrates in a direction perpendicular to the direction of motion, and propagates by small successive twists of the medium, so the medium must be able to support a shear stress. A vibrating string is a good example of a transverse wave. The parts of the string go up and down while the wave moves down the string. A sound wave, on the other hand, is a series of compressions and expansions. A loudspeaker in a stereo system responds to the varying current from the amplifier by rapidly moving in and out. During the outward motion it pushes on the air and gives it a local compression, which moves away from the speaker. During its inward motion, it sucks air inward, thereby decreasing the local density, and this rarefaction also moves away from the speaker. In each compression or rarefaction, the air molecules move only a short distance. They just go back and forth, and the result is a periodic pressure disturbance moving in the same direction as the wave.

Air can support only longitudinal waves. Steel, however, can also support a transverse wave, because when it is twisted out of its normal shape, strong elastic forces come into play that twist it back. Steel can support a shear stress; a perfect fluid cannot.[34] It can only transmit successive compressions and rarefaction along the direction of wave motion.

Newton's method of ignoring any pictorial representation of gravity, and treating it as if action-at-a-distance were a reality, had very fruitful and far-reaching results in electricity and magnetism as well as in mechanics. But understanding the ether became urgent in the latter part of the nineteenth century when research on electricity and magnetism culminated in the physical insights of Michael Faraday and in James Clerk Maxwell's definitive equations of electrodynamics.

The time and the space in Maxwell's equations were presumed to be the absolute time and space of Newton, and the ether was presumed to be the medium in which electricity and magnetism acted, and in which light was propagated. As a result, there were increasing efforts to determine the existence of the ether from

[34] A perfect fluid is one whose molecules do not interact. Air is close to being a perfect fluid because the interactions between its molecules are very weak. The molecules in water, however, interact strongly, so transverse waves are possible. In fact, much of the wave motion in the ocean is transverse.

experiment. Both continuous and particulate models of the ether were proposed.

Maxwell's work seemed to settle the question. His equations described electromagnetism by fields in which successive interactions between adjacent points in space can transmit forces over large distances. This is similar to the propagation of sound by local compressions and rarefactions.

This certainly dispensed with the troublesome idea of action-at-a-distance, but at the expense of ascribing a new function to space. Also, there was the awkward fact that Maxwell's equations worked fine for all electromagnetic phenomena on Earth. Did this mean that the ether was attached to the Earth?

Maxwell's equations described all known electric and magnetic phenomena accurately, and Newton's laws accurately described mechanics. But comparing electromagnetism and mechanics exposed a fundamental contradiction. Newton's laws were the same for all inertial systems moving at constant velocities relative to each other while Maxwell's equations were not. That is, when Newton's laws were subjected to the Galilean transformation from one coordinate system to another moving at a constant relative velocity, the exact form of the equations of motion was recovered. The second law of motion was exactly the same in all inertial systems. For Maxwell's equations, however, transforming from one coordinate system to another with a constant relative velocity *did not* yield the same result. The equations of electrodynamics were different in different coordinate systems. Mechanics obeyed the principle of Galilean relativity; electrodynamics did not. The already strong conviction that the ether had an indispensable role in physics was thereby confirmed. It was the medium, and therefore the reference system, in which Maxwell's equations were true. There is a logical disconnect here. Maxwell's equations arose from experiments that were performed in Earth-bound laboratories. That is, the data were obtained in a reference system attached to the Earth. The equations therefore had to be valid in the coordinate systems firmly fixed to Earth. If they were true only in a reference system fixed in the ether, then the ether had to be attached to the Earth. This was a remarkable requirement, implying a remarkable coincidence.

The ether had been the only alternative to action-at-a-distance, and it provided a medium for the propagation of light. Now it seemed also necessary as an absolute reference system for electrodynamics. It was an odd thing. It permeated all space, including the interatomic spaces of matter, and planetary orbits showed that matter could move through it without resistance. At the same time, the ether had to have elastic constants enormously greater than steel or any other known solid so it could carry the transverse waves of high velocity that constituted light.[35] The contradiction was obvious. Materials of greater density resist deformation more strongly and therefore have greater elastic constants. How could something that is rarefied enough to pass through the atoms of matter without being detected have such enormous strength?

In spite of this, the ether was deemed a necessity, and there seemed to be a faith that, when its properties were fully understood, there would be no contradictions.

Experimental verification of the ether's existence became ever more urgent. In particular, the ether was needed to provide an absolute space for Newton's mechanics and an absolute coordinate system for Maxwell's equations. Surely, since light consisted of waves in the ether, its measured velocity should depend on the motion of the Earth, unless, of course, the ether always moved in step with the Earth's motion.

Various experiments, including measurements of the torque on a capacitor suspended in a magnetic field, measurements of the velocity of light in moving water, and studies of bi-polarized light, were conducted to find this variation of velocity. The most conclusive experiments were those of Albert Michelson at the Case School of Applied Science and Edward Morley at Western Reserve

[35] If a material has a high elastic constant, it takes a lot of force to bend or twist it, and when it is deformed, it snaps back quickly. The response of air to a pressure is sluggish compared to that in steel, so the velocity of waves in steel is higher than that in air.

University in Cleveland of 1887.[36] These were refinements of Michelson's first experiments, published in 1881. Remarkably, the seeds of this work were personally planted by Maxwell, who fully understood the implications of his own work and had written to D. P. Todd to ask about the possibility of measuring the velocity of the Earth through the ether. Todd was an astronomer who was Director of the National Almanac Office in Washington. Michelson, who was a young naval officer at the time, had recently been transferred there, and he thought he could make the measurement. His first experiments were done in Berlin, but later he joined the faculty at Case Institute in Cleveland and teamed up with Morley, a chemist from Western Reserve, which was right next door to Case. Together, they built a very sensitive interferometer at Western Reserve and again found no velocity relative to the ether.

Michelson was widely recognized as an outstanding experimentalist in optics, having done the most careful and accurate determinations of the velocity of light. Accordingly, the results of the Michelson–Morley experiments were taken very seriously.

They used the interferometric method to detect the velocity difference when light was propagated in two different perpendicular directions. An important debate centered on whether the ether moved easily through the Earth, or whether it was dragged along with the motion of the Earth in its solar orbit. The experiments of Michelson and Morley were designed to settle this question. If the ether were independent of the Earth's motion, then their experiment would give two different values for the two directions and for different points in the Earth's orbit. But if the ether were carried along with the motion of the Earth, there would be no measurable difference. In fact they got a null result. They got the same answer for the speed of light at two points in the Earth's orbit

[36] Michelson was then at the Case School of Applied Science and Morley was at Western Reserve University. In 1947 Case changed its name to the Case Institute of Technology. The two Institutions occupied adjoining campuses in Cleveland and in 1967 they merged to form Case Western Reserve University.

that were six months apart. Since the velocities of the Earth were opposite each other at these two times, the velocity of light would be the same only if the ether were dragged along with the Earth. Conversely, if the ether were not affected by the Earth's motion, then the velocity of light would be different for experiments done six months apart.

Note that the motivation behind the experiments was not to determine whether or not the ether existed, but to find its velocity relative to the Earth. Its existence was assumed.

All attempts to determine the ether velocity gave the same result: zero. So Michelson concluded that the ether was indeed carried along with the Earth. To test this, he repeated his experiments at high altitudes, reasoning that, if the Earth indeed dragged the ether with it as it moved, the Earth's influence must surely decrease as the distance from the Earth increased. Again he found a null result. If there was an ether drag, it extended high into the atmosphere.

None of this made any sense, because it violated everything that was known about the motion of particles or of waves. If light consists of particles, then the velocity with which it reached an observer should be the sum of the velocity of light and that of the source. If, for example, a ball is thrown forward on Galileo's ship, then the observer on the ship sees the ball travel with its own velocity. But an observer on shore will see the ball moving with the sum of its own velocity and that of the ship. For corpuscles, the observed velocity will always be the sum of the velocities of the corpuscle and that of its source. The experiments on light showed that its velocity was the same no matter how the source was moving. If light consists of waves, the analysis is a little different. It is true that the velocity does not depend on the motion of the source, but it certainly depends on the motion of the observer. If a man in a stationary boat strikes the water repeatedly to set up water waves, the speed with which they move over the water does not depend on the speed of his boat. However, if a man in another boat is watching the waves, the speed that he will see depends on how fast *he* is moving. The observed velocity is the difference of that of the wave and of his boat. The ether is like the water in that it carries light waves, so observers at different speeds should see different light velocities.

No such differences were ever observed.

A number of attempts were made to explain the zero results of these experiments, the most notable being that of George Fitzgerald in Ireland (1889) and Hendrik Lorentz in the Netherlands (1892). Lorentz was one of a small number of scientists that immediately recognized Maxwell's equations as a major starting point for further work and used them to work out the theory of the motion of charged particles.

Fitzgerald and Lorentz independently proposed that there was a contraction of the length of material bodies in the direction of their motion that canceled the effect of the expected velocity difference. When this assumption was put into electrodynamic theory, everything worked out beautifully. It showed that measurements of time and distance made in one inertial coordinate system, of events in another inertial system, were altered because of the contraction. The alteration was such that no change in the velocity of light could be observed by experiments.

The ether was saved. In 1882 Lorentz had taken the ether to be at rest in absolute space and had essentially identified it with absolute space. Now he preserved this absolute ether by showing it could not be detected because of the Fitzgerald–Lorentz contraction. Nevertheless, it was recognized as an ad hoc assumption, and Henri Poincaré suggested that the impossibility of detecting motion through the ether be taken as a fundamental physical law. Poincaré was one of France's greatest mathematicians and spent much of his intellectual life at the University of Paris. Taking all of mathematics, and much of physics, as his domain, he was the first to recognize the foundation of chaos theory, namely that very small changes in physical conditions could produce large results. He worked on the Fitzgerald–Lorentz contraction and anticipated Einstein in some respects.

Yet he agreed with Lorentz that the contraction of bodies induced by motion was a real property of matter, from which the equations of what we now call special relativity follow. But it was really an ad hoc assumption, giving no insight into the origin of the contraction, its relation to measurement, or the nature of the ether. It was just *assumed* that objects got shorter when they were moving and that the amount they were shortened was just exactly enough to make any motion through the ether undetectable. It had no independent experimental basis and no function other than to save the idea of the ether. It was most unsatisfactory.

The resolution of all this was provided by Einstein in 1905. He took the null results as a fact, assumed the velocity of light was a constant in any inertial coordinate system, and then assumed that both the equations of electrodynamics and mechanics had to be the same in all inertial coordinate systems. This was the theory of special relativity. It was one of the great scientific achievements in the history of science, but it must be remembered that it had precursors, the Lorentz contraction being the most important. Also, Poincaré had thought a lot about the ether and came extremely close to special relativity. But it was Einstein who had the physical insight and the bold, independent mode of thought that provided the essential key.

In retrospect, it is surprising that it was not the Michelson–Morley results that drove Einstein to special relativity. He did not mention their work in his 1905 paper, and in later discussions he denied that they had any strong influence on him. It was an insistence on adopting inertial relativity for *all phenomena*, and his acceptance of facts as final arbiters that led to his success. The constancy of the velocity of light, or at least the impossibility of finding otherwise, was well supported by experiment even before Michelson and Morley. To this he added his conviction that there had to be a unified approach to both mechanics and electrodynamics. The addition of philosophical conviction to observation led to the special theory.

Einstein accepted the experimental results as basic facts, and if the experiments could not detect the ether, and if measurements of the velocity of light all gave the same constant value, then so be it. The special relativity theory dispensed with the ether in that it no longer had a role to play in physics. But the trade-off was that the nature of space and time was not as had been stated by Newton and implicitly accepted by all until Einstein's analysis.

The achievements of Newton and Einstein had an important commonality in that they both arose from a strict adherence to the result of experiment. But Einstein's approach was a little different. Newton was successful in consciously separating his ideas about the ether from the rigorous analysis of mechanics, which he developed on the basis of the three laws of motion. Newtonian mechanics therefore was based on experimental facts. He tried hard to understand the ether, and, rather than rejecting it, he put it aside for further study. Since there were no definitive experiments addressing its properties or its existence, this was an appropriate, conservative position. But his attitude toward the meaning of time and space was different. He apparently felt no ambiguity about these concepts; he accepted and refined the intuition that had evolved over millennia of everyday experience, thereby defining an absolute space and an absolute time.

Einstein's commitment to experiment went beyond that of Newton's and was more radical. He maintained that *only* experimentally verifiable statements should enter into a scientific description of the physical world. He was at one with Mach on this and stated that Mach's ideas had an important influence on him.

Experiments gave the result that the velocity of light was the same no matter how the source of the light or the observer was moving. In spite of the fact that this made no sense when compared to all our other experience with motion, it was an experimental fact. No experiment could measure the velocity of the ether, and the ether could not be detected in any way. The discussion of its existence is merely word play, because, as far as actual observable facts are concerned, it did not exist.

"So be it"! said Einstein. We need to take the constancy of the velocity of light as a basic reality. A more precise statement is that the speed of light in a vacuum is the same for all inertial systems.[37] The metaphysical component to Einstein's thought was essential. He believed descriptions of nature had to be internally consistent and have a basic simplicity. *All* physical laws had to be equivalent and the same for all inertial systems. It would not do for this principle to hold for mechanics and not for electrodynamics. Both sets of laws had to hold as well in one inertial system as in another.

It is interesting to look at the concepts of the ether before and after Einstein. Its most important function had been to avoid action-at-a-distance, which must be defined as the ability of one body to affect another *without anything at all* existing in the intervening space. Maxwell's equations were the ultimate way of getting rid of action-at-a-distance, because it reduced all physical effects to highly local interactions by elevating the actions of fields to a supreme position. The fields required a medium for their existence and propagation, just as for the familiar, and thoroughly understood, fields and waves in elastic media. The conceptual snag was that this medium, the ether, was thought of as a material thing and therefore had to have mass and weight. It is only a question of word usage to accept the ether as being a vacuum in which gravitational, electric, and magnetic fields exist. That is, we can simply *define* it by the observable properties of physics.[38] Of course, this means that the traditional definition of the vacuum as being something in which nothing, absolutely nothing, exists has to be modified. What we have called the vacuum we now describe as having physically real properties, and we have expanded our naïve ideas of what existence means. It used to be that we ascribed existence only to "ponderable" matter and its manifestations. That is, only objects with mass and mechanical properties exist. The

[37] This principle is often stated as "the speed of light in a vacuum is independent of its source". It is obvious that the two methods of expression are equivalent.

[38] With the advent of quantum theory, we find that the "vacuum" has a number of properties that contravene common sense, making it an ether with properties more varied and weird than anyone could imagine in the age of pre-quantum physics; but that is another story.

ether, in fact, was an attempt to define a ponderable medium to sustain the imponderable wave action of light. Now we ascribe a reality of existence to fields.

An argument can be made that all this is just a semantic exercise, and that the other redefinitions, those of space and time, were much more profound because they were not questions of language; they addressed the very foundations of scientific thought.

Yet the notion of an ether had a powerful grip on scientific thought. It had to be abandoned before any progress could be made on the road to relativity and it should have been abandoned because nothing about it could be observed: not its properties, its effects, or its very existence.

After relativity, there was no ether and no action-at-a-distance, and all physical effects propagated with the speed of light. This included gravity; there were no instantaneous interactions.

11

The genius

The history of science is filled with the stories of the great minds that have struggled to understand the physical world. Every generation from Aristotle and Archimedes to the present day has had its share of true genius exploring nature, trying to learn its secrets and bringing it into some sort of sensible order.[39] Of them all, two stand out as so much greater than the others that they can only be compared to each other. These are Isaac Newton and Albert Einstein.

Einstein has come to be the personification of ultimate scientific genius. The news of Eddington's observations in 1919 of the shift in starlight as it passed near the Sun during a solar eclipse made Einstein an instant world celebrity. *The Times* reported that a new theory of the universe had been verified, showing that space was warped and that "Newtonian ideas were overthrown". The *New York Times* followed this a few days later with headlines proclaiming that "light was all askew in the heavens", that Einstein was triumphant, and that only twelve men in the entire world could understand his work. The combination of esoteric, mysterious language, sounding like the weirdest of magic, the appeal that the stars always held, the fact that an Englishman had confirmed the theory of a German so soon after the First World War (the eclipse occurred on May 29, 1919, while the Armistice was signed on November 11, 1918), and that the news was cast in the form of

[39] Richard Feynmann called them "monster minds", intellects of such a high order that they far exceeded ordinary intelligence. He said this in his youth, not realizing that he was one of them.

a conflict with Newton, not only the greatest scientist that ever lived, but also an Englishman, was overpowering.

Only later was it realized that the Eddington's observations did not really prove that Einstein was right. The definitive proof had to await more accurate measurements. This made no difference to the public perception of Einstein; he already was, and remained, the greatest scientist in the world.

The idea that relativity was too difficult even for most scientists is a myth. Its basis is very simple and the mathematics it uses is well known. Relativity is based on one single experimental fact and one single idea. The experimental fact is that the velocity of light in a vacuum is always a constant, and the idea is that *all* the laws of physics are always and everywhere the same.[40] This is relativity's entire foundation. Of course, a lot of work is needed to work out the full consequences of this, but it is all within the knowledge base of competent theoretical physicists. The mathematics for special relativity is quite elementary and is taught to undergraduates in physics and mathematics courses. General relativity involves a more complex mathematics, called tensor analysis, which was not a part of university physics curricula at that time. But much of its complexity arises from the detailed notation needed to express it, not from anything intrinsically hard in its concepts. It has been worked out over a period of time, and anyone capable of doing theoretical physics could readily learn it. The obstacles to the mathematics of general relativity yield to the same kind of patient effort needed for other branches of science. The difficulty in relativity was in conceiving it, not in learning it. Furthermore, the ideas and consequences of both special and general relativity can be understood without using any advanced math at all and can be made accessible to any intelligent person. It is not the inherent content that makes relativity hard to

[40] Actually, the laws of physics could not be the same everywhere unless the speed of light were a constant, because otherwise Maxwell's equations would be different in different inertial systems. Thus there really is just one basic idea. The laws of nature are the same everywhere for everyone.

understand; it is the necessity to throw out preconceived concepts that are unconscious parts of our everyday lives. Once we get beyond these prejudices, the rest is just ordinary logic. Of course, as for all scientific subjects, the logic must be tight and rigorous and requires disciplined concentration, but there is nothing in the theory that cannot be understood.

Einstein's fame increased with the years, and his image of genius solidified and grew, particularly after he moved to Princeton and acquired a giant halo of white hair along with a reputation for gentle humanity and for being wise about all things, not only science.

A comparison of Einstein with Newton is inescapable. They both completely transformed the science of their day. They both addressed the deepest issues of scientific knowledge; they both spent a lot of effort working on light, gravitation, and motion; they both immeasurably enhanced our understanding not only of terrestrial phenomena, but of the entire universe. The parallels are transparent and obvious. Both of them had a huge capacity for work and both ascribed their successes to ordinary human qualities, Newton stating that the major difference between him and other men was that he would stay with a problem constantly and indefinitely until he was satisfied with the results, and Einstein maintaining that he was successful because he was extremely curious and could not let go of a scientific issue until he understood it.

Yet they were completely different in background, temperament, behavior, and world outlook. Their differences began with birth, since Einstein was a healthy baby while Newton was so small that his survival was in doubt. And Einstein's family life was completely different from Newton's. The Einsteins were a caring family, and, although the father, Hermann, was often on the verge of bankruptcy, the family was always able to provide for Albert's needs and education, which they did ungrudgingly with the help of Albert's wealthy maternal grandfather. Unlike Newton, Einstein did not have to fight his family about his interests. On the contrary, he was immersed in an environment that valued the intellectual life. It is true that Hermann wanted Albert to study something "practical", especially electrical engineering, so he could help in the family business, and it is true that his mother strongly opposed his first marriage, but Albert managed to go his own way while retaining his family's love and support. Of course there was conflict. Einstein was something of a rebel, opposing much in the character of academe and the intellectual life of the times because it was narrow and overly nationalistic. He often opposed existing norms and beliefs. His family, on the other hand, was quite conventional, committed to middle-class aspirations, and they wished he would be a conventional professor with a conventional life. This was not to be, but there never were fatal rifts.

A major difference between the two was their approach to knowledge. Einstein believed that observation, experiment, and reason were the only roads to acquiring knowledge about the physical world while Newton saw physical science as one part of an all encompassing truth that included alchemy, philosophy, and religion. Newton was deeply religious, while Einstein was raised in a freethinking, liberal Jewish family and, although he did go through a religious phase in his youth and ultimately became an active Zionist, was more of a pantheist than anything else. However, this does not mean that Einstein's work was devoid of any philosophical content. On the contrary, his work was based on an insistence that experimental, and observational results could be understood only from the point of view of philosophical premises. One of these was that nature was basically simple, and relativity arose from his conviction that the fundamental laws of physics had to be the same everywhere and at all times. That this had to be true for inertial systems gave him special relativity. That it had to also be true for systems in accelerated motion gave general relativity. There was no a priori basis for relativity to be universally valid except philosophical conviction.

But these differences must not obscure the fact that their philosophies were, at bottom, identical. They both believed that there was a fundamental unity in the world and they both believed it could be found by reason, observation, and experiment. Their difference was in religion. Einstein was, at most, a pantheist, while Newton believed in an active God, present in the world and essential to it. His belief was passionate and unquestioning, but *his road to knowledge of nature did not include personal revelation or religious faith.* It was always based on facts and reason.

Einstein had an artistic bent that was completely alien to Newton. At his mother's persevering insistence, he learned to play the violin at an early age. Contrary to his expectations, he came to love music and became a good violinist.[41] Less widely known is that, largely through his own efforts, he also became a pianist, and often relaxed by playing and improvising on the piano.

Newton spent almost his entire life at Cambridge, leaving only to take a government appointment in London. He was a true loner with no societal involvement, other than with the Mint, and had a rather testy, suspicious relationship with the scientists of his time, jealously fighting to claim priority for his scientific and mathematical discoveries. Einstein, on the other hand, was a world traveler, lived in several different countries, and was heavily engaged with many other scientists, with whom he had cordial and mutually respectful relations. In fact, he published jointly with over thirty other physicists. Also, he was a kind of activist, lending his name to various humanitarian causes, espousing pacifist and anti-war views, and supporting Zionism through writings and fund raising.

They were complete opposites with respect to their sexuality. Newton never married and most probably never had a close relationship with any woman, while Einstein was married twice and fathered three children. The first was the illegitimate daughter of Mileva Maric, whom he later married. Einstein never saw that daughter, whose name was Lieserl, and in spite of the best

[41] I recently met someone whose aunt was a concert pianist and sometimes played duets with Einstein. She thought Einstein was a rather poor violinist. The judgment of a concert pianist is bound to be different from that of an adoring family and friends.

efforts of historians, what became of her is still uncertain. Einstein thoroughly enjoyed the company of women, was intimate with his second wife Elsa even before he divorced Mileva, and had several extra marital affairs.

Newton was quite famous during his lifetime and was recognized as the world's greatest scientist, but Einstein's fame was of a different kind. From Europe and the Americas, to India, China, and Japan, he was regarded by the general population, as well as by many intellectuals, as the greatest scientific mind that ever lived. His name came to explicitly denote super genius and was used as an adjective and a noun to describe the highest degree of intelligence. There seems to be no decline in the awe he continually inspires, and the number of papers and books written about him and his work is still growing.

His image, and the mystery attached to his theories, was created by the mass media, but his reputation as a legendary genius is well merited. His popular fame rests on his theories of relativity, but he addressed other important scientific questions with an equally powerful intellect. As early as 1902, at the age of 23, he published papers on thermodynamics, electrochemical potentials, and intermolecular forces. While these did not shake the foundations, they represented sound scientific work and were a taste of things to come.

The year 1905 has been labeled the "Annus Mirabilis", Einstein's year of miracles. It is comparable to the two plague years Newton spent at Woolsthorpe. Einstein's doctoral thesis of that year was on the sizes of molecules, an innovative exercise, since many of the great scientists of the time had reservations of the utility of the concept of molecules and atoms, and some even doubted their very existence. Led by Wilhelm Ostwald at Leipzig, the Energetics School believed that the atomic hypothesis would be unverifiable and not useful, preferring to base science on a thermodynamic-like approach in which the basic concept was energy, not matter. Einstein's thesis was good work and verified the existence of atoms, but it was the other four contributions that made 1905 so memorable: the papers on the photoelectric effect, Brownian motion, special relativity, and the mass–energy equivalence.

When light shines on a metal, electrons are ejected from the surface. A number of experiments had been done to study this photoelectric effect by shining monochromatic light onto a metal surface. The number of electrons emitted and their energies were measured as a function of the intensity and frequency of the light striking the metal.

The results were simple, but hard to understand. For a given metal, no electrons at all were given off if the frequency of the light was too low. Once a threshold frequency was exceeded, the number of electrons emitted was proportional to the intensity of the light, but the energy of each electron was *totally independent of the intensity of the light and depended only on its frequency*. The experiments could be extrapolated to very weak light beams, and there was no doubt that even when the electrons left the metal one at a time, they all had the same energy. Increasing the intensity of the light only increased the *number* of electrons, not their energy. This was a mystery, because light was a wave, and the greater the intensity of the light, the greater the amplitude of the wave and the greater its energy. Therefore, more intense light should result in more energetic electrons. Experiment said no. The electron's energy could be increased only by increasing the *frequency* of the incident light. The strangeness of this result is evident from thinking of a light wave striking the metal surface. How could a wave transfer its energy to a single electron? Did the metal somehow store that energy until there was enough and then focus it onto the location of the electron, thereby kicking it out of the metal? Imagine an ocean wave rolling up a beach and imagine it all coming together to concentrate on a pebble, pick it up and kick it into the air! It looked as if this was what light did to the electrons in a metal. And why should the frequency have anything to do with it? The energy of a wave was in its amplitude; the greater the amplitude of a sound wave, the louder the sound, and the higher a tidal wave, the greater its destructive force. Light was different. Experiment showed that its intensity had no effect on the energy of individual electrons. Increased intensity just increased the number of electrons ejected from the metal. For the individual electrons only the frequency mattered.

Einstein knew of Planck's work five years earlier on black-body radiation.[42] When an object is heated, it gives off light. At low temperatures, it gets warm, but no visible light is given off. As the temperature of the body is raised, it first starts to glow cherry red, then yellow, and then white. Questions of fact arise. How much light is given off? What are the frequencies of the light and how are they related to the temperature? A straightforward controlled experiment is needed to give unambiguous answers. The experimental apparatus consists of a hollow box or sphere with a small hole in its surface and surrounded by a heat source. As the box is heated, the radiation inside it can be examined by looking into the small hole. After the box has been heated to some temperature and held there for a time, equilibrium is established. That is, there are no changes with time in the frequencies of any light escaping through the hole. The cavity is filled with radiation, because heat excites the atoms in the sphere. The electric charges in the vibrating atoms move and generate electromagnetic waves, as described by Maxwell's theory. The waves bounce around inside the cavity and can be absorbed by its atoms. A balance between the emission and absorption of radiation by the atoms keeps the cavity in equilibrium.

The experiment consists of measuring all the frequencies of the light emitted through the hole and the intensity for each frequency. Spectroscopic equipment is available to do this. The cavity is called a black body because it absorbs all the radiation falling on it. (An important part of the experiment is to make the hole so small that the amount of light that gets through it is a very small fraction of the total radiation inside the sphere.) None of the light seen by

[42] A glowing object is said to be a black body if no radiation falling on it is reflected. All of it is absorbed and then re-emitted by the object. Graphite is the ordinary material closest to being a black body.

the experiment is reflected from the walls of the cavity. It is all absorbed and re-emitted.

The term "black body" was coined in 1862 by Gustav Kirchoff at the University of Heidelberg when he showed that the radiation in a black-body cavity depended only on temperature and was independent of the material forming the cavity. He showed that if this were not so, heat could be made to flow between two black bodies at the same temperature, thereby violating thermodynamics.

The first point to note is that the total amount of energy emitted per unit time from a black body is proportional to the fourth power of the temperature. This is required by thermodynamics and must be correct. More interesting data comes from examining the intensity of waves of various frequencies. The experiment shows that, for a given temperature, the intensity of waves of very low frequency is low and increases with increasing frequencies until it reaches a maximum and then, at very high frequencies, decreases to zero. The data created a crisis because it was contrary to the theory of the time. Well-established, classical electrodynamic theory predicted that the energy of the radiation in a black body should continually increase with increasing frequency, so that ultraviolet waves carried much more energy than infrared waves. Classical theory also stated that *all* frequencies are possible and that every frequency contributes to the energy of the black body. This was known as the ultraviolet catastrophe, because, when the energy of all the frequencies was added up, using the rules of classical physics, the amount of energy in the black body was computed to be infinite! Obviously this could not be.

The experimental data were in direct contradiction to the classical theory of the time, which took light to be a continuous energy flow. It was a wave and all frequencies of electromagnetic radiation were allowed, each frequency carrying an amount of energy depending on its intensity. Thus, a wave of a single frequency could have a varying amount of energy; the greater the intensity, the greater the energy. The classical theory added all these together and got the wrong answer: infinity! It's hard to imagine a more drastic disagreement with experiment.

Planck made sense of the data by taking the energy of a light wave to consist of discrete packets, which he called quanta. He assumed that the energy of each packet was proportional to the frequency of the light wave, not its intensity, and got an equation that agreed with experiment. He was most disturbed by his own work, because he thought it upset an important part of the world's essential harmony. Nevertheless, he reported his results, because he was a great scientist who went where experiment and logic led him. Planck's formula for the distribution of radiation had far-reaching results for astronomy and cosmology. The details of the distribution depend on the temperature of the radiating body, so that the temperature of a star can be computed from spectroscopic observations of its radiation. Also, a major piece of evidence for the Big Bang theory of the origin of the universe is the low temperature of the radiation permeating all space.

Einstein was the first to take the Planck hypothesis literally, saying, in effect, that if light were a packet, then it would travel through space like a packet and in fact could be regarded as a particle with a certain energy. The particle is called a photon. The energy of a light beam was not determined by the amplitude (intensity) of the light wave, but by its frequency. More energetic beams just contained more photons, each of which had the same energy as all others for light of a given frequency

When Einstein applied this idea to the photoelectric effect, he found that all the experimental results could be reproduced if the energy of that particle was proportional to the frequency of the light. *Furthermore, the data showed that the constant of proportionality for the photoelectric effect was the same as in Planck's formula for black-body radiation.*[43] The interpretation of the data was then straightforward. A particle of light entered the metal, collided

[43] This is the famous Planck constant, which appears in all of quantum theory, from the uncertainty principle to the energy levels of atoms and the properties of nuclei.

with an electron, and kicked it out. The energy of this particle of light, the photon, was the same as the energy associated with a given frequency in Planck's radiation formula. Even the fact that there was no photoelectric emission until some critical minimum frequency was exceeded could be explained by this theory. At too low a frequency, the photon did not have enough energy to get the electron out of the attraction of the interior of the metal. A "light particle" had to have enough energy, that is a high enough frequency, to give the electron enough energy to kick the electron through the metal surface.

In a single paper, Einstein had shown that the wave theory of light, which had been standard science since shortly after the time of Newton, was inadequate, and he thereby established the validity of the quantum hypothesis. The importance of this paper cannot be overstated. Conventional wisdom dates the beginning of quantum theory to the work of Planck, but Einstein's theory of the photoelectric effect was the first explicit statement since Newton that light was like a stream of particles, and in fact the first time that the wave–particle duality of light was recognized,[44] because, while Einstein treated light as particles, he still retained its wavelike character, since he said its energy was determined by its frequency, a purely wave concept. His photoelectric paper was a major step to the quantum theory of matter and light. Yet his ambivalence about quantum mechanics showed up even then, right at its origins, since he included the word "heuristic" in the title, suggesting that his approach was a guide to the true theory, not the final word.

The microscope and telescope opened new worlds of the small and the large to scientific analysis. One of the most important microscopic discoveries for physics was that of the motion of small particles suspended in a liquid. In 1827, Robert Brown, a Scottish botanist working in England, looked at pollen grains suspended

[44] This may not be quite true. In trying to describe the nature of light, Newton certainly adopted the corpuscular view. But he also introduced some wavelike concepts.

in water under a microscope and found that the grains were in constant motion. The motion was random, each particle moving erratically from its original position, taking short, arbitrary jumps, but never moving very far from its original spot. By the time Einstein became interested in this phenomenon, it was already known that the motion was less rapid the heavier the particles, and the idea had been proposed that the jumps were the result of particles being randomly bombarded by the water molecules. The important element in Einstein's analysis was that the appropriate variable to look at is the mean square displacement.[45] Many people knew that the essence of Brownian motion was the motion of a particle by a series of random, independent jumps. That is, if a particle is subject to a random force, it will move a short distance in some arbitrary direction, all directions being equally probable. The particle is said to execute a random walk.[46] Clearly, the particle will not wander far from its original position in a given time. Being random, the jumps have as much chance of bringing the particle back to where it started as to some new position. The scientific issue is to describe the motion in such a way as to give the probability of the particle's position at a given time.

An illustrative example is the "staggering drunk" problem. An individual leaves a bar after a long drinking bout and tries to find his way home. He starts by walking along the sidewalk until he reaches an intersection. Being well oiled, he has no idea of where he is, or what direction he should take. There are four possible directions and he just picks one and goes on. The chance of picking one of the directions is the same for all of them, and he is in no condition to make a sensible choice. The chance of taking a particular direction

[45] Each time the particle jumps, measure the distance it travels and square it. Take the average of these squares over very many jumps. This is the mean square displacement.

[46] The correct designation would be "random flight", because the particle moves in three dimensions. But so many examples, especially for teaching purposes, are two dimensional, that the term "random walk" is widely, if inaccurately, used.

is the same for all four, namely 25 percent. He walks until he gets to another intersection and again has a 25 percent chance of going in any one of four directions. One important question is: in any given amount of time, what are the chances that he will be some given distance from the bar? Another is: what are the chances that he will get home after some given time has passed, and what is the most probable time it will take him to get home? This simple, ludicrous example has many important analogues and extensions in the physical world.

Previous attempts to understand Brownian motion were based on considerations of the velocities of the particles. These led to a dead end. Einstein fully recognized the true probabilistic nature of the process and united it with a molecular collision mechanism, thereby starting a line of study that embraced a great variety of physically important processes. The theory of the diffusion of matter through gases, liquids, and solids, sedimentation, colloids, coagulation, fluctuations, random noise in electrical circuits, the gravitational effects of random distributions of stars, even the conformation of long polymer chains: all arose from the concepts and mathematics in Einstein's Brownian motion paper. With his doctoral thesis and a follow-on paper in 1906, he established the relation between fluctuations and diffusion, worked out a method of counting the number of molecules (Avogadro's number), and estimated the sizes of molecules.

It was a brilliant extension of the statistical mechanical methods of Boltzmann and Maxwell and an exposition of important probabilistic consequences of the existence of molecules. Its prediction of the average distances suspended particles would travel per minute was verified by the French physicist Jean Perrin. He began his experiments in 1908 using improved microscopes to follow the detailed motion of suspended particles. His work completely verified Einstein's theory.

In one stroke Einstein verified the existence of molecules, along with their sizes, rationalized what had heretofore been a set of puzzling observations, and laid the foundations for an important and still growing body of theory.

While not recognized at the time, Brownian motion and the photon were both important in the processes of stellar evolution and the processes that maintained a star's size. In a normal star, the temperature is high and photons are continually being emitted and absorbed by ionized atoms. Radiation from the interior of the star could not simply pass through as a wave and escape from the star. Rather, light, as photons, acts as if being bounced around just like the pollen grains under Brown's microscope and takes a long time to get to the star's surface. Photons help fight the compressive force of self-gravitation, and without their Brownian motion, a star would cool off very quickly and could not last for billions of years.

The work of 1905 was important, and it was the remarkable study of the photoelectric effect that was cited as the basis for awarding Einstein the Nobel Prize in 1921. This work alone would have marked him as one of the century's great physicists. But he accomplished even more, and created even more profound revolutions, with his two papers on special relativity, also published in that year of miracles. The first paper resolved the most pressing difficulty of the time on the foundations of science. The sciences of mechanics and electrodynamics gave different pictures of the world, not because they dealt with different phenomena, but because they treated space differently. For mechanics, different coordinate systems moving relative to each other at constant velocity obeyed the same physics. No inertial coordinate system was better than any other. Thus the laws of mechanics were the same for someone standing on Earth as for someone riding in a train. Electrodynamics, however, seemed to imply that some special system was preferred to any other, and this was identified as being attached to the ether. Experimental investigation failed to find this ether and the conflict of the two sciences, and the paradoxical results of experiments, created an impasse more serious than anything since the time of Newton. Einstein resolved this by taking the boldest step imaginable, accepting the seemingly ridiculous notion that the velocity of light was the same for any velocity of its source or of the observer. He ignored the ether and required only that the laws of physics be the same in all systems moving relative to each other, both for mechanics and electrodynamics. This led him to a radical change in our ideas of time and space: — so radical that they were shocking to his contemporaries and still seem weird to us. The paper containing the special theory of relativity was the start of a new era in science. Then, in a follow-up paper, he used special relativity to show that matter and energy were two aspects of the same thing and that matter could be converted to energy. The profound implications of this result became evident with the development of the atomic bomb in the Second World War. Never before has a purely theoretical scientific discovery had such enormous global significance.

Scientists, as well as the public, have called Einstein one of the greatest minds of all time. Yet he had many detractors, and he attracts strong criticism to this day. The criticism is of two kinds: that coming from people with little sound scientific knowledge and from those with a particular personal antipathy to Einstein. Sometimes these were ideologically motivated, as were the venomous attacks by some Germans during the Nazi regime, and sometimes by people who simply did not like either, his fame or the results of his research. An entire book has been written contending that Einstein was the supreme plagiarist, stealing other people's work and calling it his own.

The Nazi criticisms were particularly virulent because they strongly emphasized that part of the German intellectual tradition which held German culture to be unique, different from, and superior to that of other peoples. The claim carried over to science, which was thought to be culture bound, and, since German culture was superior and right, "Jewish physics" had to be wrong. This attitude foreshadowed the thesis that all knowledge, including science and mathematics, was only conditionally correct, depending entirely on the milieu in which it was embedded. In its modern form, it is claimed that there is no absolute truth. It is all relative to culture and is really about power relationships. Cultural relativism reached its heights in American universities during the seventies and eighties and had profound negative implications for Western intellectual life.

Another class of critics consisted of knowledgeable scientists, historians, and mathematicians, who were particularly critical of his special relativity on the grounds that it was anticipated, and practically completed, by others. Whittaker, a distinguished mathematician who wrote a detailed *History of the Theories of Aether and Electricity* credited others with special relativity, saying that "in 1905 Einstein published a paper which set forth the relativity theory of Lorentz and Poincaré with some simplifications". When told of this by Max Born, Einstein replied: "If he manages to convince others, that is their own affair. I myself have certainly found satisfaction in my efforts, but I would not consider it sensible to defend the results of my own work as being my own 'property,' as some old miser might defend the few coppers he laboriously scraped together". Can you imagine Newton answering an attack on his priorities in such a manner?

Fitzgerald had indeed proposed the contraction of measuring rods when they moved rapidly; Lorentz had invented the correct transformation equations for converting space and time between coordinate systems moving with respect to each other; and Poincaré had proposed relativity schemes like Einstein's. Physicists at that time were fully aware of this work, and of other attempts to make sense of the consequences of the constancy of the velocity of light and of the inability to detect the ether. Einstein knew at least some of them personally. He particularly admired Lorentz, whom he visited several times, and regarded him as the grand old man of physics. In his later years, Einstein said that special relativity was ripe for discovery, implying that if he had not done it, someone else would have.

But the physics community gave the greatest credit and recognition to Einstein. There were sound reasons for this. It was Einstein who insisted on taking the experiments as the full truth, used them as postulates to form physical theory, and took the results for the nature of space and time as inescapably true. He accepted the far-reaching changes that special relativity demanded in our conception of the most basic components of the world. The second reason is that he developed the relativistic idea into a complete theory, embracing all of mechanics and electrodynamics. His anticipators, while men of great intellect and physical insight, neither created the revolution nor carried the new vision forward into all branches of physics. Even after Einstein's 1905 paper, Lorentz found it hard to give up the ether. Precursors and anticipation always exist for any great shift in scientific paradigms. In the case of relativity, it was Einstein who crystallized the known science into a consistent whole and transformed it into a new world-view.

A different kind of criticism was that in a book published by a Serbian countrywoman of Einstein's first wife that claimed it was Mileva who first raised questions about the ether and initiated the work that became the theory of relativity and that she and Einstein were coworkers in its creation. It was claimed that everything Einstein created later stemmed from the work of this husband–wife team. There is no good evidence for any of these claims, and the author seems to have been motivated by nationalistic pride.

A dominant factor driving the acceptance of the theory of relativity is its beauty. It has the major characteristics of great classical art: elegance, balance, contrast, point of focus, tension and resolution, thematic inevitability, and an overall compelling harmony fusing its parts into a unified whole. Its aesthetic appeal stems from its inherent simplicity, which is attractive in itself, and simply awesome when viewed in the light of the great structure of so much varied, yet connected, results flowing from a very few, very simple assertions. The artistic sense as a motivating factor in scientific research was well expressed by Poincaré: "A scientist worthy of the name, above all a mathematician, experiences in his work the same impression as an artist. His pleasure is as great and of the same nature". Poincare stressed the artistic attractions of mathematics because he was, above all else, a mathematician. But the same is true for all great scientific work, experimental as well as theoretical.

The constancy of the velocity of light, the principle that gravitational and inertial mass are equivalent, and the invariance of physical laws are enough to construct a theory that spans the cosmos. Only the great symphonies can compare to the theory of relativity as works of art. Yet even they do not have its simplicity of basic material, its enormous reach and complexity, and its powerful thematic impact.

Einstein's antipathy to quantum theory is well known, and his debates with Niels Bohr are legendary classics in which two giants of twentieth-century physics struggled over the meaning of the new results stemming from the study of atoms and subatomic particles.

Bohr took the results as valid and embraced a probabilistic picture informed and enhanced by his complementarity principle. The results of experiment were indeed probabilistic, and indeed fundamental entities sometimes exhibited wave-like and sometimes particle-like properties. These were not two mutually exclusive pictures of nature; they were complementary. Which one was dominant was determined by the particular experiment being performed. Reality was *not* independent of the nature of the observations being made.

Einstein could not accept this. He maintained that there was an objective truth and a strict causality that was independent of the observer or the nature of the observations. He repeatedly presented Bohr with "thought experiments" that purported to show that quantum mechanics could not be the final word, and Bohr repeatedly showed that the thought experiment overlooked something or was being misinterpreted and that quantum theory was correct after all. It was a truly deep and fundamental issue, bearing on the very nature of reality and the possibilities of scientific knowledge, as well as on the status of the theory of matter.

The crucial point of the debate was documented in a 1935 publication by Einstein, Podolsky, and Rosen, the content of which came to be known as the EPR paradox. It distilled the essence of Einstein's views. The authors considered thought experiments in which two particles, initially at rest and at the same place, were given equal and opposite velocities, so that both their future positions and momenta had to be opposite to each other. Now they assumed that a measurement of position could be made on one of the particles. This could be done to any degree of precision desired; it was only the *simultaneous* measurement of momentum and position whose precision was limited by quantum mechanics. Since the other particle had the opposite position, they knew its position with complete accuracy, *without ever having disturbed it by a measurement.* The same could be done for a measurement of momentum. The two particles were very far apart and therefore could not affect each other. All properties of the particle therefore had to have an objective reality, and objective values, that had nothing to do with the possibilities of measurement. The quantum theory must therefore be incomplete since it predicts that the uncertainty principle holds under all conditions. For this to be true in the EPR experiment, there had to be some instantaneous connection by which one particle could affect another instantaneously over enormous distances, which Einstein called "spooky action-at-a-distance".

Experiments were later carried out in which this spooky action-at-a-distance was actually observed! The crazy consequences that Einstein took as proof that quantum mechanics could not be a complete theory were actually seen! The meaning of this is still a subject of debate.

To the end of his life, Einstein believed that quantum mechanics was an approximation to an objective, causal theory that would explain and resolve the issues raised by its probabilistic nature.

Yet he continued to make important contributions to quantum theory. In 1906, he had worked on the theory of the absorption and emission of light by atoms and in 1917 published a major paper on the spontaneous and stimulated emission of light that was the foundation of laser action.

In 1907, he resolved the great specific heat paradox in which experimental data showed that the du Long–Petit rule, which said that the heat capacity per atom was the same for all solids, was wrong, except for high temperatures. Classical theory gave this incorrect result and Einstein showed that, by applying quantum theory to the atomic vibrations in solids, correct results were obtained for all temperatures. This was the beginning of the vibrational theory of the thermodynamic properties of solids, which has grown into a major part of modern solid-state theory.

In 1925, Einstein's interest in the statistical mechanics of quantum gases was sparked by some work of Bose, who showed that Planck's results for the radiation field of a black body could be described as a gas of particles of zero mass, called photons. In the second of three papers on gases, Einstein applied similar ideas to material particles, showing that the classical theory of ideal gases was valid only in the high-temperature limit. He developed the theory of what is known as the Bose–Einstein statistics and of the Bose–Einstein condensation, which is now the subject of so much experimental research.

This was not all. At that time, Louis de Broglie, in a splendid leap of imagination, stated that, just as light had corpuscular properties, so material particles should have wave-like properties. De Broglie's thesis advisor, Langevin, asked Einstein's advice on the suitability of this subject for a PhD thesis, and Einstein quickly replied in the affirmative. In fact, this was the basis for Einstein's argument that, if particles had wave properties, then they should follow the same statistical laws as photons.

All this would be considered as a corpus of magnificent work on quantum mechanics by anyone. In spite of his distaste for quantum mechanics, Einstein went where the facts and logic led him.

As early as 1907, while trying to fit gravitation into special relativity, he realized that special relativity needed to be extended and had "the happiest thought of my life". He realized that someone falling from a roof would feel no gravitational field. It was a simple idea, but it led him to conclude that accelerations were equivalent to gravitational fields and this led him to the general theory of relativity. In 1911 he published a paper on gravity and light, which was the beginning of his final ideas on general relativity. His major paper on the subject appeared in 1916, and once again he stunned the world with his radical approach to the nature of light, matter, space, and time. The creation of general relativity was a long, extremely intense process, often frustrating and always exhausting. Only a mind like Einstein's could have carried it to completion. Two years later, he presented a bonus to the scientific world by applying general relativity to the universe as a whole, thereby founding the modern theory of cosmology.

An interesting point is that David Hilbert presented the theory of general relativity to the Royal Academy in Gottingen five days before Einstein presented it at the Prussian Academy in Berlin. Hilbert was one of the greatest mathematicians of all time. A major objective of his was to put all of mathematics on one unified axiomatic basis, an objective that was never achieved. His impact on modern theoretical physics, however, cannot be overstated. His ideas permeate all modern methods of analysis, from differential equations to quantum theory. He had read Einstein's papers and had been intensely interested in physics for several years, so he invited Einstein to visit him. Einstein stayed a week, gave a number of lectures, and was pleased that his work was thoroughly understood and accepted. On the basis of what he learned from Einstein, Hilbert constructed the equations of general relativity. He got them before Einstein because he was a much better mathematician, and Einstein kept making mistakes that had to be corrected before arriving at the final results. Hilbert gave full credit to Einstein as the inspiration and originator of general relativity, pointing out that there were many people who were much more competent at differential geometry than Einstein, but only Einstein's incredible physical insight and firm convictions could have created the theory.

Einstein never liked quantum theory, so he did not contribute much to either its wave or matrix formalism, nor to many of its applications to atoms and molecules. Instead, he spent the rest of his life trying to generalize his work to combine electric, magnetic, and gravitational forces into a single theory. Some have deplored this as being fruitless, wishing that Einstein had instead accepted quantum mechanics and applied his great talents to it. I disagree. Only someone of his abilities had any hope of making any progress whatever in uniting electrodynamics and gravity, and being well established, of maturing years, and highly respected, he had nothing to lose and perhaps much to gain for physics by pursuing a unified field theory. The search still goes on, although along a different track than that taken by Einstein. The modern efforts are not based on pure space-time continuum theory, but try to combine quantum mechanics and relativity to account for nuclear, as well as electrodynamic and gravitational, forces. The most promising current approaches postulate the existence of tiny strings in multiple dimensions that are curled up so that at the macroscopic level we see only three of them (plus time). The various manifestations of particle forces and particle masses are thought to be associated with the vibrations of these strings and their uncurling. The jury is still out, but we must remember that it was Einstein who first dreamed of a "theory of everything".

A remarkable fact is that Einstein did the work for his year of miracles while a part-time scientist. He held a job as a clerk in the Swiss patent office from 1902 to 1909, when he resigned to become a professor at the University of Zurich. He did not neglect his work at the patent office during this time, and in fact had been promoted in 1906. But the duties were not onerous, and he spent every extra waking hour on physics.

Widespread media attention and public acclaim is surely not a sufficient reason to label Einstein a towering genius. His critics point out that special relativity had been anticipated in many important respects by predecessors and contemporaries, and there were many who were better and more subtle mathematicians. In fact, the proper four-dimensional mathematics for special relativity was produced by Minkowski, not Einstein. And the mathematics needed for general relativity was developed by a series of mathematicians from Gauss to Ricci and Levi-Civita. Einstein was not fond of mathematics for its own sake; he seems to have regarded it as a necessary labor that had to be done to get at the golden nuggets of truth about nature. His passion was physics, not mathematics, and he sometimes got others to do the long chains of mathematical analysis he needed for his work.

Media perception does not often reflect reality, but they certainly agree in the case of Einstein. My reasons for stating this go to the meaning of scientific genius.

That he was not a pre-eminent mathematician has nothing to do with his singular position in the history of science, because his work was in physics, not mathematics. Please note that he was able to quickly learn and apply all the math he needed, no matter how advanced or complex.

He created a revolutionary, brand new paradigm for understanding the physical world. It is true that there were others who understood the transformation equations of special relativity before Einstein, but only he took the incredibly bold step of following experimental facts on the velocity of light to a redefinition of space and time, the two most fundamental descriptors of reality.

Great new science is always a beginning; it inspires further important work, spreads out to encompass ever growing numbers of natural events, and continually enhances understanding of the physical world. In this regard, Einstein's work is comparable only to that of Newton. Keep in mind that great consequences arose from his work on the statistics of small particles and quantum mechanics, as well as from special and general relativity.

Yes, he did not accept quantum theory. However, he rejected it only because he could not believe that it was a final theory. He fully accepted the facts of quantum experiments and the results of the current quantum theories, and he contributed some of the most important ideas to the field. But he regarded the theories as provisional, expecting that they would be replaced by something grander and more in accord with his belief in a rational universe. This is quite a different matter than simply rejecting quantum mechanics.

By any measure of changing a world-view, clarifying the most fundamental concepts, and starting the most significant research over a long period of time, Einstein was at the very summit of scientific genius.

12

Time and space

All motion, from planetary orbits to a ball on a pool table, takes place in space over a period of time. Time and space seem to be among those aspects of the world that cannot be analyzed beyond our innate sense of their meaning. They are more than just a part of our existence; they practically define it, and they are built into the human psyche. We all know what we mean by "before and after" or "here and there". It doesn't seem possible to go much deeper than our compelling inner sense of time and space, which is a clear and dominant aspect of consciousness.

Yet, when we use them in physical theories that aspire to a degree of precision, as in Newtonian mechanics, we feel bound to analyze them as well as we can so that some agreed upon sense can be applied when they appear in equations or in descriptions of experiments. Newton's discussions of space and time illustrate the problems that arise when attempts are made to quantify obvious intuition. We can do no better than refer directly to the *Principia*.

Newton starts with a list of seven definitions to set up the language of his mechanics and immediately follows this with a scholium (explanatory note). It is a long explanation, going on for nearly seven pages in the Cohen-Whitman translation, but its essence is contained in the first few paragraphs:

1. Absolute, true and mathematical time, in and of itself and of its own nature, without reference to anything external, flows uniformly and by another name is called duration. Relative, apparent, and common time is any sensible and external measure of duration by means of

motions: such a measure—for example an hour, a day, a month, a year—is commonly used instead of true time.

2. Absolute space, of its own nature without reference to anything external, always remains homogeneous and immovable. Relative space is any movable measure or dimension of this absolute space…[and] is determined by our senses from the situation of the space with respect to bodies…

From our modern perspective, these two statements tell us nothing except our inner feelings about time and space. In fact, the entire scholium is just a refinement of our sense that there is an absolute time going continually forward that is other than, and independent of, any measure by any sort of clock, and there is an absolute space other than, and independent of, any sort of measurement, such as that of the distance measured by a ruler.

Absolute time agrees with our internal experience. We automatically accept the idea that events have an order we call "before" and "after" and that, although our sense of duration depends very much on circumstances, beneath it all there is a flow that is steady and unvarying. The idea that there can be a fundamental variation in this flow has no meaning. Furthermore, there is no way to imagine that time can "start" or "stop", so the extension of absolute time into the infinite past and infinite future is in accord with our intuitive sense.

Time is measured with the aid of clocks, which are either counting devices or mechanisms for converting measures of time into measures of distance. We recognize that clocks can give different values for a time interval, but ascribe this to the imperfections inherent in any mechanical device. In spite of different clock readings, we insist that time intervals are the same everywhere. When clocks differ, we synchronize them by resetting them to be equal to each other or to some other clock deemed to be more accurate. Until recent times, celestial motions were used to provide such a clock on the assumption that they were much less prone to variation than the mechanical clocks we build on Earth. Currently, the ultimate time standard is given by measurement of the frequency of radiation emitted by cesium atoms. This frequency depends only on the energy states of the atom and is independent of any external influences, and therefore provides an extreme degree of precision.

But Newton believed that underneath it all is that unstoppable, unchangeable flow of universal, absolute time that has been going on forever and will continue forever. Everyone believed this until Einstein showed otherwise.

Time is homogeneous. A physical process requiring a certain interval of time will require that same interval, no matter when or where it is measured, provided only that all physical conditions are the same. Thus, the time it takes for a ball to fall through a given distance (in a vacuum) is the same now as it was at any time in the past, or will be at any time in the future, and it will be the same anywhere else, if the gravitational field is the same. There is a fundamental assumption here, namely that identical physical processes take place over identical time intervals. When Galileo measured the swings of a church chandelier with his pulse beats to conclude that each swing took the same amount of time, his tacit assumption was that the interval between the beats was always the same. And in using his water clock to measure the time it took for a ball to roll down an inclined plane, he assumed that the time interval for equal amounts of water took the same time to leave the hole in the bottom of the barrel. Using more accurate clocks makes the same kind of assumption. The count of cycles in the frequency of an atomic clock is taken to be a measure of time, and this assumes that the time of a cycle is an invariant interval. There is no way of verifying this experimentally, and in fact it doesn't even make sense to look for such verification. The *assumption* of the homogeneity of time makes good sense from a physical point of view, because if we choose one clock as a standard, we can see that identical process, when measured by some other clock, indeed do take identical times. But the definition of "identical processes under identical circumstances" must be complete. Not only must the physical mechanism and the processes they measure be the same, along with any environmental factors such as fields, temperature, and but humidity, but they should also be as close together as possible and not be moving relative to each other.

An important consequence of the homogeneity is that, given identical clocks, all observers will note identical times for a given event. Newton believed this to be universally true and not only for clocks that were close by and stationary with respect to each other. In particular, he took the measurement of time to be independent of place *and* of the motion of the clocks. Time was the same for someone standing still as for someone moving rapidly, the same on Earth as on the Moon or on a distant star. His concept of absolute time did not recognize its homogeneity as a definition, but took it as a fundamental property of nature.

Newton's absolute space also extends to infinity, but in three directions rather than one because space is three dimensional. Being absolute, it has the same properties at all places and in all directions. That is, absolute space is homogeneous and isotropic. Again, he took these attributes to be fundamental properties of space rather than either definitions or propositions to be demonstrated by experiment.

What we actually measure, however, is not absolute time, but readings on clocks; and not absolute space, but distances on rulers or measuring rods. Newton recognized that neither of these is absolute, but this did not disturb his belief in the existence of absolute time and space.

Distance certainly depends on the coordinate system from which we make our measurement. The ball dropped from the top of the mast on Galileo's ship falls straight down when observed on the ship, so a measurement along the deck will yield a zero displacement. But an observer on the shore will see a displacement determined by the forward motion of the ship and therefore will note a finite displacement when he uses a measuring stick attached to the shore. Newton says that both systems, ship and shore, are relative spaces, that the connection between them via their relative velocity is well understood, and that the laws of motion in both spaces are the same when measured in each of their coordinate systems. This connection is simple. To transform from one inertial system to another, it was only necessary to add the relative velocity between the two. This is the Galilean transformation.

The measurement of time also depends on our frame of reference. The moment at which the ball hits the deck of Galileo's ship is easily recorded if there is an accurate clock on board. But consider an observer on the shore, who has an identical clock and also records the moment of impact. He will not record the same time because he must get a signal that starts from the point of impact, and this takes time to reach the shore. The observer on shore will therefore record a later time than the observer on the ship. The Newtonians knew this and accommodated it in two ways: first, the commonly used signal was light, whose speed was so high that it did not enter into the usual terrestrial mechanics. For most practical purposes, the speed of light could be taken as infinite and the clocks would indeed read the same time on the shore and on the ship. While this did not work for interplanetary or interstellar distances, clocks at different places could be made to give identical readings by correcting them using the known velocity of light. This is the result of the following thought experiment. Assume that two clocks are synchronized and one is placed on Mars while the other remains on Earth. Now send a light beam to Mars that reflects off the Martian clock and returns to an observer near the Earth clock. The observer sees that the reading from Mars is not the same as on the Earth clock. The Earthman sees a reading on the Mars clock that is earlier than that on his clock by the time it took for the reflected light to reach him. The calculation is easy because he knows the velocity of light and the distance to Mars. The Mars clock is made to give the same time as the Earth clock by simply adding the calculated time it takes light to travel from Mars. The Newtonian view holds that by correcting for the time it takes for light to reach us from some other object, no matter where it is, or how fast it is moving, we get a time that is always and everywhere the same.

Maxwell's work emphasized that the light velocity had to be measured relative to the absolute ether, and the failure to detect any ether velocity led to Einstein's revolutionary insight that destroyed the belief in an absolute, universal time.

There was nothing intrinsically complicated about Einstein's solution to the impasse forced on physics at the end of the nineteenth century. He started from the constancy of the velocity of light in a vacuum. The experimental evidence for this was overwhelming, so Einstein proposed that we take this as a fundamental fact and forget the ether. All attempts to explain this fact ended up with results in which the ether played no role and could not be detected experimentally, so it doesn't count. An ultimate statement of the most important principle of physical science is this: only that which is experimentally observed, or that which can be logically connected to experimental observation, has any reality. This was not a revolutionary idea. It only seemed so when applied to light because the universal constancy of its velocity, totally independent of the motion of the light emitter or the light receiver, is counterintuitive and doesn't seem to make sense.

Einstein just added to this his requirement that there be a certain unity and symmetry in nature. Mechanics was the most successful and complete part of science ever developed, and its laws were known to be the same in all inertial coordinate systems. Electrodynamics clearly had a range and importance comparable to that of mechanics, yet it laws were *not* the same in all inertial systems. When the laws were transformed from one system to another moving with a constant velocity, the form of Maxwell's equations changed. Einstein found this to be intolerable. The experiments on light showed that there was no ether, so another approach to rationalize electrodynamics was needed. He simply *assumed* that nature had to be simple and unified, so, if the laws of mechanics were the same in all inertial systems, it had to be true that the laws of electrodynamics were also the same in all inertial systems.

Once the idea that the ether provides an absolute frame of reference was abandoned, these two statements had to be accepted. Applying logic and mathematics to them gave the special theory of relativity.

The surprise was in the results. By straightforward logic, it followed that the traditional notions of time and space had to be modified. Until special relativity, it was implicitly assumed that simultaneity had a specific and absolute meaning. That is, if two different events, such as switching on two flashlights, were seen to take place at the same time when measured in some coordinate system, they would be seen to take place at the same time when measured in any other coordinate system. Two simultaneous events were thought to be simultaneous in all inertial systems. This would certainly be true if the signals by which we looked at clocks in different systems had infinite velocity. It would also be true if the signals had a finite velocity, provided the measured velocity of the signal had the velocity of the source or the observer added to it, because then we could use this additivity to unequivocally determine the times of the events. This is true for sound waves. If we are moving through air and a bell attached to the Earth clangs at some instant, we can tell if the clang occurs at the same time as a bell moving with us simply by adding our velocity to that of sound when we calculate how long it took for the clang of the Earth-bound bell to reach us.

Light waves are different, and it is light waves that are important. Light is the fastest signal in the physical world, and its most basic property, that the velocity of light does not depend on the velocity of its source or of the observer, gives it a universality lacking for any other signal. Furthermore, all physical effects, including electricity, magnetism, and gravity, propagate with the speed of light. Even if we use sound waves, as in the above example, we determine that we are moving relative to the Earth by using light. The velocity of light enters intrinsically in all physical phenomena.

The constancy of the velocity of light is a peculiar thing and violates our sense of the nature of motion. But it is an inescapable experimental fact, and Einstein's genius lay in accepting this fact, and accepting it completely.

And this led to results that seem strange even today. All the intuitively obvious notions of space and time hold quite well, as long as all measurements are made within a given inertial coordinate system. But if an observer in one system measures space and time intervals for events in another inertial system, the results seem odd indeed. If an observer in one system, say attached to the Earth, makes measurements of what is happening in another system, say attached to a train or a plane, he finds that *lengths are contracted and time is slowed down*. That is, if the Earth-bound observer measures the length of a rod moving relative to the Earth, he finds that its length is less than that of an identical rod that is stationary on the Earth. Also, if he measures a time interval using a moving clock, he finds it to be longer than that for a corresponding interval recorded on a clock in his stationary system (the Earth). Moreover, if an observer in the moving plane or train makes measurements, he finds exactly the same thing. To him, the length of a rod is shortened and clocks are slowed down in the Earth-bound system.

In summary, the lengths of moving objects are shorter and moving clocks are slower in all inertial coordinate systems. This is just the Fitzgerald–Lorentz contraction, but it was not introduced in an ad hoc manner to save the concept of the ether, as in previous work. It was rigorously derived from the experimentally demonstrated fact of the constancy of the velocity of light! There was no intrinsic property of matter that compensated for the failure to find a light velocity relative to the ether. Rather, the constant velocity showed that the nature of time and space itself was such that the distance contraction and time dilation for moving rods and clocks was the reality.

It sounds weird, but its origin is very simple if we remember that length is what we measure with rulers and time is what we read on clocks. Read this again and keep it in mind as you try to understand the effects of motion on time and distance. In physics, length is defined as the result of a measurement, such as laying a ruler along a table edge, and space is defined as the totality of all such possible measurements. In fact, no presupposed concept enters into this. The distance between two points is just the number we get out of the measuring process. Similarly, a time interval between two events is the result of a measurement by a clock. It has nothing to do with what we might feel time to be or how we think it flows in our consciousness. “Time” and “space” are only about measurement by instruments. Nothing else.

There are many thought experiments that have been constructed to illustrate that, because of the finite velocity of light, measurements of time and distance will give different results for observers moving relative to each other. Too often, these seem to be a source of perplexity, rather than clarification. To choose clarity over confusion, the reader needs to accept two points. The first is that the thinking in the thought experiment, while just ordinary logic, requires close attention and cannot be absorbed without some concentrated work. But there is nothing extraordinary about this: it is the same kind of effort needed to understand the proof of a theorem in elementary geometry. The second point to remember is that we are talking about *physical* distance and *physical* time as measured by rulers and clocks. I am repeating this because I believe that the greatest obstacle to clarity is the stubborn subconscious tendency of our minds to carry our subjective feeling of time into the analysis. The result of using these physical devices for measurement *must* lead to results that are different for moving bodies than for those at rest. Since light takes time to reach us, the time intervals we see registered on a moving clock cannot be the same as on a stationary one. The light path from the clock to us is different at the beginning and at the end of the time interval. The same goes for the reading of rulers. The role of a finite speed of light was recognized in pre-relativity physics. The difference in relativity is that it takes the speed of light in a vacuum to be always and everywhere the same. This is the origin of the weirdness. But again, it is worth repeating that the constancy of the speed of light must be accepted, because it is an undeniable physical fact.

As an example of a typical thought experiment, let's construct a simple device by placing two mirrors one above the other, a fixed, vertical distance apart. Let's place a light source at the bottom mirror such that a beam of light moves from the bottom to the top mirror. Just attach two small mirrors at some angle to the two larger mirrors so that a portion of the light ray leaving the bottom mirror and a portion leaving the top mirror are reflected to our clock. Our clock will register a time reading when it receives the first signal (time of emission from bottom mirror) and a later time reading when it gets the second signal (reception of light from the top mirror). This is the time difference between the two events: emission and reception. These measurements are made with the mirror apparatus being stationary with respect to our clock. What happens if we set the apparatus in motion parallel to the ground and again measure the time of arrival from the bottom and top mirror? For simplicity, assume that when the bottom mirror emits its light beam, it is in the same position as that of the bottom mirror when the apparatus was stationary. Then the time it takes the light from the bottom mirror to reach our clock is the same for the moving and the stationary clock. But the top mirror is another story. Because the mirrors move while their light flashes travel towards us, the top flash takes longer to reach us than it did when the mirrors were stationary. The time difference we measure between emission and reception is therefore *longer* when the clock is moving. Clearly, we can calculate the difference between the two cases from some simple geometry because we know the velocity of light, which is always the same. This is a general result. Motion dilates time intervals of events in a moving system.

It is worthwhile looking at another experiment, because it is so simple and informative. Imagine watching a plane fly overhead (and imagine you can see through it) in which a light flash is set off exactly midway between two passengers, one at the front and the other at the rear of the plane. Assume they have clocks that have been synchronized. These two observers will record exactly the same time on their two clocks because light takes the same time to go the two distances. The observer at the midpoint of the plane sees the light signals arrive at the front and rear at the same time on his clock. That is, the two events, arrival at the front and arrival at the rear, are simultaneous. But what about the observer on the ground? He sees the arrival of the light to the front passenger *after* he sees its arrival to the rear passenger, because the front passenger is moving away from the flash while the rear passenger is moving towards it. That is, the two events, arrival at the front and arrival at the rear, are *not* simultaneous when looked at from the Earth. Using light to measure the distance between the front and rear passengers similarly shows that a measurement taken by an observer inside the plane will not give the same result as a measurement taken by someone on the ground. *The results of measurements of time or distance depend on the states of motion of the observers.*

It is clear that the conversion of the space coordinates and the time from one inertial system to another has to be different than the simple Galilean transformation, which was implicitly based on the assumption that the speed of light was infinite. Einstein sought the transformations that arose from the constancy of the speed of light and found the Lorentz transformation.

There is nothing difficult or complicated about these results; they follow immediately from the act of taking a measurement on a ruler or reading a clock. They address the same issue Galileo faced when he looked at the motion of an object on a moving ship and found its velocity as seen from the shore was just the velocity of the object as seen from the ship plus the velocity of the ship relative to the shore. But there is a critical difference. Galileo implicitly assumed that distances and time intervals were the same for all moving systems. Einstein showed that this was incorrect.

Derivation of the Lorentz transformation equations requires only elementary plane geometry and elementary algebra. But they completely changed our intuitive notions of space and time, which were embodied in Newtonian mechanics. The effect is small at velocities much smaller than that of light, but becomes dominant as the velocity increases. Even at a velocity as high as ten percent that of light (18,600 miles per second) the ruler contracts by less than one percent. The contraction increases rapidly at high velocities and the ruler shrinks to half its original size when the velocity is about eighty-seven percent that of light.

The time dilation is equally small at low velocities and equally dramatic at high velocities. If its velocity is ten percent of the velocity of light, a moving clock reads only five percent slower than a corresponding stationary clock. But when the velocity gets to three quarters that of light, the time dilation is such that the moving clock is going only half as fast as a stationary clock. At ordinary terrestrial velocities, the relativistic effects are incredibly tiny. Even for the rotational velocity of the surface of the Earth, which is about 108,000 kilometers per hour, they are about one part in a hundred million. This is why Newtonian mechanics has been so successful.

An important property of special relativity is that measurements of time and distance are not independent. In Newtonian mechanics, the measurement of distance has nothing to do with time. It is the same process and gives the same result at any time. Likewise the measurement of time has nothing to do with the position of the clock. In special relativity, however, the distance measured in one inertial system depends on *both* the time and the distance measured in another. Similarly, the time read on a clock in one system depends on both the time and distance read in another. Time and distance are inextricably linked. They are not independent entities. That this is so is evident from our thought experiments given above and is embodied in Minkowski's famous statement that neither space nor time have an independent reality. Minkowski's mathematics gave rise to the idea of time as a fourth dimension.

Of course, time is not a fourth dimension in the spatial sense. We call a line one dimensional because only one number is needed to locate a point on it, and we call a plane two dimensional because two numbers are required to label a point. These two numbers can be the x and y coordinates on axes of the ordinary Cartesian system of plane geometry. If a third coordinate axis is erected perpendicular to the other two, we have a three-dimensional system which needs three numbers to identify the point. These three numbers specify length, breadth, and depth. It is nonsense to think we can put in a fourth axis to give a four-dimensional space. This is not the meaning of the "fourth dimension" in relativity. But note: mathematically, the essence of dimensionality is *the number of parameters needed to identify a point.* In our ordinary space the number of parameters is three. Mathematically, a higher "space" can be defined by requiring more than three parameters to label a "point". In special relativity an "event" is defined as a point in space at a given time. The event takes four numbers to identify it: three space coordinates to locate it, and another number to specify the time at which we look at that space point. It is in this sense that we relate time to a "fourth dimension".

Actually, the description of time as a fourth dimension is a little deeper than just requiring four numbers to label an event. To see this, recall how distance is described in ordinary three-dimensional geometry, in which the distance between two points is obtained from the Pythagorean theorem. Assume our distance starts at the origin of our coordinate system and consider the x, y, and z components of the line from the origin to the end of our distance. Then, from elementary geometry, the square of our distance is just the sum of the squares of the components. In special relativity, it turns out that a simple quantity related to time[47] combines with the squares of the space coordinates to give a sum that follows a Pythagorean theorem completely analogous to that in the ordinary space of three dimensions. The only difference is that it has four squared quantities in it instead of three. Since the three-space is Euclidean, the analogous four-space is also called Euclidean. An equivalent statement is that the space-time continuum is flat.

This analogy is important because it allows all the results of ordinary geometry to be generalized to space-time. The meaning of the fourth dimension is a mathematical meaning, but it has far-reaching physical implications because it explicitly recognizes the fact that time and distance measurements cannot be separated from each other. Treating physical phenomena as events in a four-dimensional space-time continuum displays the inseparable connection between time and space, and leads to methods of

[47] This quantity is just the negative of the product of the velocity of light squared and the time squared.

treating the quantities of mechanics in terms of four-dimensional quantities.

All of the mechanics of special relativity follows rigorously from the Lorentz transformation, including the mass–energy equation. This is the most famous physics formula in the world and states that the two seemingly distinct entities energy and mass are really one and the same, and are related by

$$E = mc^2.$$

E is the energy of a body of mass m, and c is the velocity of light. Since the velocity of light is so huge, it takes only a small amount of mass to give an enormous amount of energy. This equation has been verified in a number of ways, most notably by the explosion of nuclear bombs, in which vast amounts of energy are released from relatively small changes in the masses of atomic nuclei. Observation shows that the energy produced by the annihilation of colliding electron–positron[48] pairs is related to their initial masses precisely in accord with this equation, so that the equivalence of mass and energy applies to a body's total mass, as well as to the small changes in mass involved in nuclear reactions.

Einstein originally arrived at his mass–energy equation by considering the amount of energy in a sphere enclosing an object that emits light waves. Because of the relativistic contraction, this sphere becomes an ellipse when the object is moving, and encloses a different amount of energy. The radiant energy for the moving body and the mass–energy equation follows directly from this result.

A more general derivation starts from Newton's second law, which states that the force acting on a body is its rate of change of momentum. The momentum depends on the mass and the velocity of the body. In classical mechanics, in fact, it is just their product.

[48] A positron has exactly the same mass as an electron but it carries a positive charge. It is therefore attracted to an electron, and when the two meet, they vanish in a burst of energy.

The situation is not quite so simple in special relativity. Because of the Lorentz transformation, the velocity, being just the rate of change of distance, is different when measured in two inertial coordinate systems moving relatively to each other. We cannot then expect the momentum to be so simply related to the force. However, the relativistic force law must reduce to the Newtonian result when the velocity of the body is small, so we take the relativistic momentum to be a function of mass and velocity. There are a number of ways of getting at the relativistic momentum. One way is to look at the collision between two bodies and another way is to examine the transformation of a particle's velocity measured in two inertial systems moving with respect to each other. They all give the same answer. The relativistic momentum is still the product of the body's mass and velocity, but the mass increases with the velocity. This means that the inertia of the body is greater when the body is in motion. The increment of force needed to accelerate a moving body is greater the faster it is moving. This is not as counterintuitive as time dilation or distance contraction, but it does sound a little strange to those who have been taught that the mass of a body is always a constant.

The direct relationship between momentum and kinetic energy leads to the mass–energy equivalence formula, and since different forms of energy can be converted into one another, the mass–energy relation is universal.

It is easy to show that momentum is still conserved in special relativity, but the laws of conservation of energy and mass must be combined to give a law in which the *sum* of mass and energy and its mass equivalent is conserved, although neither may be conserved individually.

Newton's foresight when he defined force as producing momentum changes, rather than just velocity changes, makes his second law still valid in the theory of special relativity.

Note that we now have an alternative physical description of the fact that no velocity can exceed that of light in a vacuum. If we try to increase the velocity of a moving object by adding another velocity to it, then we must add a second Lorentz transformation to it. The algebra shows that no matter how many new velocities we add, the velocity of the object approaches a limit that it can never exceed: the velocity of light. The velocity dependence of the mass means that as the velocity of an object increases, its mass goes up and an ever increasing force is needed to accelerate it further. The force approaches infinity as the velocity approaches the speed of light, which, therefore, can never be reached.

The relationship between the mass and velocity is just like that between time and velocity. The mass of a body moving relative to our laboratory coordinate system increases with velocity in exactly the same proportion as a clock attached to that body slows down. Just as for time dilation, therefore, the mass of a moving body is nearly constant at low velocities and increases dramatically at high velocities.

The energy content of mass is enormous, as shown by the existence of nuclear bombs. But the amount of mass converted to energy in a nuclear bomb is very small. If a single gram of mass were completely converted to energy, the result would be equivalent to twenty-one-and-a-half thousand tons of TNT.

Special relativity completely transformed our concepts of space and time because it dealt only with what we actually do to *measure* space and time, not with what we feel about space and time. The measurements are objective; the feelings have been conditioned by millennia of experience with everyday distances and time intervals.

13

It really is true

It's worthwhile restating some of the results of special relativity theory, because they are so counterintuitive:

1. A moving object is shortened in the direction of its motion.
2. A moving clock runs more slowly than one that is stationary.
3. A moving object is heavier than the same object when stationary.
4. Energy and mass are the same thing.

These results are so peculiar that a great amount of effort has been expended in devising seeming paradoxes that arise from the theory and then trying to resolve them. The most famous of these is the "twin paradox". One of two twins takes off in a rocket ship, goes to a distant star, turns around, and comes back to find that he is much younger than the twin he left behind. Because of time dilation, the time spent on the trip, as experienced by the twin in the rocket, is much less than the time that has passed for the stationary twin. But, the paradox goes, since motion is relative and the traveling twin can see himself as stationary while the Earth-bound twin is moving, he must see the Earth twin's clock as moving more slowly, so that it is the Earth-bound twin that is younger. When they face each other, both cannot be the younger. Something is wrong. Many papers have been written on this. The resolution of the paradox is that the motion of the two twins is not symmetric. The second twin must change velocity twice: on leaving Earth and on turning around to get back. Special relativity applies only to constant velocity, symmetrically moving coordinate systems such that if one is moving relative to another with a given velocity, then the first is moving with respect to the other with the negative of that velocity. Since the second twin undergoes acceleration, special relativity is inadequate to analyze the relative aging of our twins. General relativity is required. In fact, Einstein published a paper in 1918 describing the twin problem by general relativity and found that indeed the traveling twin came back much younger than the one who stayed home. There is no paradox when the right theory is used.

Another paradox is that of a pole moving through a barn with doors at its front and back ends. Let the pole, when stationary with respect to the barn, have the same length as the distance between the front and back doors, so it can fit inside the barn. Now give the pole a constant velocity parallel to the ground toward the front door. We see that the pole has shortened because of the Lorentz contraction, so it is an easy matter to close both barn doors when the pole is completely inside the barn. But if we sit on the pole, we see the barn moving, and the Lorentz contraction causes the barn to become shorter, so the pole can never fit inside. From one point of view, the pole can fit inside the barn; from another point of view it cannot.

It is easy to show that this is not really a paradox, because of the phenomenon of time dilation. The two times, those at which the front and back end of the pole enter the barn door, are different for the two observers, and a simple calculation shows there is no paradox. The pole can fit into the barn for both observers. This paradox arises from an incomplete application of special relativity and does not need general relativity for its resolution.

The literature on relativistic paradoxes is enormous. The universal result is that they can always be straightened out by the correct application of relativity theory.

Straightforward logic and mathematics shows that the theory is internally consistent. That is, if the two basic postulates, that the velocity of light is independent of the motion of the light source, and that the laws of physics in all inertial systems must be the same, then the theory has no internal contradictions. In itself, this does not show that the theory is right. Only experiments and observations can decide if the theory is correct.

The fundamental assumptions are true. Every experiment shows that the speed of light (in a vacuum) is always the same, no matter how the light source is moving and no matter the direction. And the evidence that the fundamental laws of physics do not depend on the state of motion is overwhelming. Einstein noted the fact that the physical result of a current being produced by a magnet and a conductor does not depend on which is considered stationary. From the point of view of a stationary conductor and a moving magnet, an electric field is induced at the magnet. If the magnet is stationary and the conductor moving, there is no electric field near the magnet, but a magnetic field is created which induces an electric field in the conductor. In both cases, the current in the conductor is the same and depends only on the relative velocity of the magnet and conductor.

The two postulates of relativity are well founded. How about their consequences? In fact, there is a massive amount of data that demonstrate the truth of special relativity, and no data that contradict it. The demonstrations of the mass–energy relation from nuclear reactions and from electron–positron annihilation have already been mentioned. An even more direct experiment has been performed, by B. Bertozzi in 1964, who used a linear accelerator to speed up electrons to speeds very close to that of light and then dumped them into a calorimeter, thereby directly measuring the heat energy they produced. He found that this heat energy corresponded to the electrons' kinetic energy just as predicted by the mass–energy equation. Furthermore, the experiments showed that the speed of the electrons never exceeded that of light, no matter how much energy was pumped into them. This experiment was a direct demonstration that mass and energy are equivalent, as well as that the velocity of a material particle could never exceed that of light.

The time dilation effect has been directly observed in two ways. In 1938, Ives and Stillwell looked at the radiation emitted by moving hydrogen atoms. The radiation displayed a Doppler effect, because the atoms were in motion. When a moving source emits waves, the frequency of the waves changes. If the waves are coming towards the observer, their frequency goes up, whereas if they are moving away, the frequency goes down. The reason is simple. If the waves are approaching, the crest of a particular wave is closer to the crest of a previously emitted wave because of the motion of the emitting source. The wavelength (the distance between crests) is therefore smaller and the frequency is higher. If the source is moving away, then the distance between crests is larger and the frequency lower. This is why a driver being overtaken by an ambulance hears the siren increase in pitch as it comes up to him, and then decrease in pitch as it passes. It was first described mathematically by Doppler.

Light is a wave motion and therefore exhibits a Doppler effect. However, when the phenomenon is analyzed according to relativistic, rather than classical, theory, the Doppler effect is smaller because of time dilation. The observer sees a longer time between two approaching wave crests, and therefore the wave has a lower frequency than if there were no time dilation. This opposes the rise in frequency of the approaching wave, so the effect is less than the classical result. The Ives–Stillwell experiments gave the relativistic result, thereby confirming the existence of the dilation.

An even more direct demonstration is that of the decay of muons, as first shown by Rossi and Hall in 1941, and later by Frisch and Smith to a greater accuracy (1963). Muons arise from cosmic rays and bombard the Earth in large amounts. They are particles with a mass about two hundred times that of an electron and a speed close to that of light in the upper atmosphere. They are not stable and decay into electrons and neutrinos[49] with a half-life[50] of 1.52×10^{-6} seconds, as measured by a clock moving with the muon (proper time).

[49] Neutrinos were first predicted by Pauli to be produced in certain nuclear reactions, and their existence is now well established. They are hard to detect because they have no charge and little, if any, mass.

[50] The half-life is the time for the decay to decrease the number of particles by one half. For stationary muons the half-life is measured by looking at the rate of decay of muons that have collided with atoms and therefore been brought to a very low velocity.

As muons move towards Earth from a great height, their number decreases exponentially at a rate controlled by the half-life. From the half-life, it is easy to compute the proper time it takes for the muons to travel from the mountaintop to sea level, as would be measured by a clock moving with the muons.[51]

In the experiment, measurements of the number of muons were taken on Mt. Washington in New Hampshire, at a height of 6,265 feet. The low-altitude measurement was taken at 10 feet above sea level. Because the velocity of the muons is known (about ninety-nine percent that of light), the time it takes for the muons to get down to sea level, as measured by a clock attached to the Earth, is known. Comparing these two times showed that the muon time, as seen by an Earth observer, runs at only one-ninth the speed of the corresponding Earth clock, in complete agreement with the time dilation formula of special relativity. A more convincing demonstration of time dilation can hardly be imagined.

The consequences of special relativity have all been confirmed, not only in these experiments, but also in many others. We should have expected this because they follow from the two basic postulates with unassailable logic. Note that there is nothing in the theory that refers to the nature of light or the constitution of matter. Yet when some information about light, such as that it is a wave motion, or on material constitution, such as the decay of muons or nuclear reactions or electron–positron annihilation, is involved in the experiments, the theory gives the right answers.

Einstein's genius lay in his insistence that only what we actually observe can enter into a description of the physical world. Other scientists had stated and believed this, but Einstein held to it rigorously and completely, and worked out all its consequences no matter what preconceived ideas it contradicted. A correct description of how we actually measure time and space is essential. Once

[51] The proper time is that measured by a clock in a coordinate system attached to it. That is, it is the time on a clock attached to the moving object. The clock on my kitchen wall shows the proper time for my kitchen.

this is carefully sorted out, relativity theory follows. The only physical fact in the theory is that light, whatever it is, moves in a vacuum with the same speed no matter how or where it is measured. This is where the weirdness comes from. But there it is. All experiments confirm the constancy of the speed of light. It is interesting to recall that the theory of electrodynamics would have some serious problems if the speed of light were not constant, because, in the Maxwell equations, the speed of light is just the ratio of electrostatic to electromagnetic units. If the speed of light were different in a space ship, say, than on Earth, this ratio would have to be different. But this is hard to accept because the measurements that define the units are done by manipulating actual objects in coordinate systems that are stationary with respect to those objects. The relativity principle for the laws of mechanics therefore implies the relativity principle for all physical laws, because all measurements are ultimately reducible to mechanical operations.

If we are to construct our view of the world from actual observation, then the strange results of relativity must be accepted.

14

The space–time continuum

Between any two points, there can always be found another point. This is a fact from ordinary high-school geometry. Put two dots on a piece of paper. Then another dot can be placed between them. Of course, if the dots are too close together, they will overlap, because the pencil marks have a finite size. But geometry is an idealized subject in which points have no dimension and can never overlap, so no matter how close together, two points always have other points between them. This is what is meant by a continuum. A line is a one-dimensional continuum with an infinite number of points between any two given points. Similarly, a surface and a volume are two- and three-dimensional continua respectively. The term continuum simply means that there are no gaps between the points.

In our daily lives we live in a three-dimensional space continuum subject to the rules of ordinary Euclidean geometry, and it works, provided it is supplemented by another continuum: that of time. Everything we do and every process we see take place in time, which we assume is a one-dimensional continuum such that between any two instants of time, there are an infinite number of other instants.

Space and time are both infinitely divisible. All the points in space are joined seamlessly together, and the instants of time flow smoothly forward.

Electrodynamics, elasticity, matter and heat flow, the theory of fluids, astronomy, and Newtonian mechanics have all achieved a high state of development using the idea of objects in the space continuum moving regularly through the time continuum. This concept has served us well, but it does have limitations.

There is a difference between mathematical geometry and physical geometry. Euclidean geometry starts from a set of axioms about the properties of points and lines from which all geometric statements follow by rigorous logic. There is no arguing with it; once the axioms are given, the rest is unavoidable. It is a self-contained deductive system. The axioms certainly grew from human experience, but once adopted as completely general, they are just a set of arbitrary definitions. The true essence of mathematical geometry is the logical development of consequences from the postulates.

Recognizing this raises the possibility of creating geometries that are different than Euclid's. In fact, there is a long history of concern about the Euclidean system, because of the parallel postulate, which states that, through any point outside a given line, one and only one line can be constructed parallel to that line. That is, there is only one line through the point that will not intersect the given line within *any* distance, no matter how large. The parallel postulate has a different character than the other Euclidean axioms because it deals with an indefinitely extended line, so it is a statement about infinity. Note that for a finite distance, there are an infinite number of lines going through a point that do not intersect the given line. The parallel postulate has meaning only for an infinite line.

Geometers were worried about this as early as the fourth century BC when they unsuccessfully tried to define away the parallel postulate. Much later, attempts were made to prove that the parallel postulate was correct by assuming its opposite and deriving absurd conclusions. This led Saccheri, in 1697, to a great many conclusions he thought were nonsense, and Lambert, in 1766, continued this work. But neither of them recognized that they were working out a non-Euclidean geometry. The first mathematician to recognize the independence of the parallel postulate from the other axioms was Gauss, and he certainly understood the possibility of non-Euclidean geometries. But he did not publish his work, because Euclidean geometry was thought to be absolutely true. Immanuel Kant himself had asserted that it was an inevitable necessity of thought. It was Bolyai and Lobachevsky in the 1820s who conclusively demonstrated that complete, self-consistent geometries could be constructed that were different them Euclid's. The ultimate development of this line of thought was differential geometry, which was a generalization of Gauss' theory of surfaces and defined a geometry by the properties of the possible expressions for the distance between two points that are very close together. Riemann brought this idea to a high state of development, and it was his work that provided the basis for the mathematical development of general relativity.

The importance of this for mathematics, and for physics, is immense. Mathematical geometry is based on arbitrary postulates; it is not an unquestionable fact of nature. Physics is based on measurements using physical objects, such as measuring rods. Whether or not these measurements follow the rules of Euclidean geometry is an open question to be settled by experiment. Physical geometry need not be identical to a particular mathematical geometry. Still, no matter what geometry is adopted, space is described as a continuum.

There is a different kind of issue we must contend with here. Special relativity shows that every measurement of distance involves the time, and every measurement of time involves the distance. If measurements of distance (space) are made relative to one inertial coordinate system, then moving to a different inertial system requires that we move from one set of coordinates to another. But in doing this, the measurements of time are unavoidably changed. That is, measurements of the space continuum in two different sets of space coordinates are accompanied by a change in the measurement of the time continuum. The space continuum and the time continuum are intimately linked. Minkowski's original quote is still the best expression of this link: "Henceforth, space by itself and time by itself are doomed to fade away into mere shadows, and only a kind of union of the two will preserve an independent reality".[52]

It is possible to still think of physical objects as existing in a space continuum and moving through a time continuum, but this picture is forced and does not arise naturally from special relativity.

The natural picture is that of a continuum that includes both space and time. Instead of a point being in space and moving through time, it is more natural to regard it as being given by four numbers, three for its spatial position and one for time. These four numbers define a point in the four-dimensional space-time continuum. It is called a world point because it locates events in both space and time in the physical world. To be more precise, consider some physical happening, such as a flash of light. Of course, such a flash takes place in a spatial region over some period of time, but we can idealize it by taking the spatial region to be arbitrarily small and the time duration arbitrarily short. (This is the same kind of idealization we use to define a material particle.) The flash is an event at a point in the space-time continuum. A material particle persists in time, so it is therefore said to consist of a "world line" in the space-time continuum.

The space-time continuum is described by a four-dimensional geometry. It has been a fruitful concept for working out the consequences of special relativity theory, as well as forming a base for the formulation of general relativity. But some care must be taken in interpreting the meaning of this four-dimensional geometry, because the time dimension is nothing like the three space dimensions. A space dimension, for example, can be rotated into another space dimension as when the x and y coordinates of a Cartesian system are rotated about the z-axis. The time dimension, by itself, cannot be rotated about anything. But a general four-dimensional "rotation" can be defined that simultaneously involves space and time in analogy with rotations in ordinary three-space. These

[52] This statement is in the first paragraph of Minkowski's 1908 paper.

"rotations" turn out to be the Lorentz transformations, showing how useful the geometric analogies can be.

Let's take a look at ordinary three-dimensional Euclidean geometry as studied in high school. The fundamental geometric concept is that of the distance between two points. If we construct a Cartesian coordinate system using three mutually perpendicular coordinate axes, any point can be labeled by three numbers, one for each axis. A point labeled by the three numbers (1, 3, 5), for example, is located at a position found by starting at the origin of the coordinate system, measuring off one unit along the positive x-axis, then moving three units in the positive y direction and then moving five units in the positive z direction. The three numbers (1, 3, 5) are the (x, y, z) components of the line joining the origin to the point, and the distance from the origin to the point is readily calculated from the Pythagorean theorem. This, in fact, is the rule for measuring distance in Euclidean geometry.

It is not a universal rule. For example, the shortest distance between two points on the surface of a sphere is the segment of a great circle, and the distance between these points is not given by the Pythagorean theorem of Euclidean geometry. A different rule, readily obtained from the geometry of a sphere, is needed. The rule for calculating distance is called the metric.

It is often convenient, and even necessary, to consider two points that are very close together with very small differences between their components, and the metric is usually given by the formula for calculating infinitesimal distances. The Pythagorean theorem for a very small distance is called the metric of the Euclidean three-space.[53] All the properties of the three-dimensional space follow from the metric. It is called the metric because it is the essential component for making measurements.

Mathematicians love to generalize, and the idea of a three-dimensional Euclidean metric is easily expanded by simply

[53] It is the fact that the infinitesimal square of the distance can be written as a sum of the infinitesimal squares of the components that makes our space Euclidean. If it is impossible to do this, the space is non-Euclidean.

increasing the number of terms in the Pythagorean theorem. If a fourth term is added, then, by definition, we get the metric for a four-dimensional Euclidean "space", which is a convenient shorthand for four sets of numbers, each set being ordered just like the numbers along ordinary three-space axes. They are said to form a four-dimensional continuum. The fact that this "four-space" is not like the three-space of our perceptions doesn't matter. After all, geometric three-space is itself just an abstraction from real, physical space. What matters is that a whole set of properties and relations follow from the four-metric that are completely analogous to those of three dimensions, giving a complete four-dimensional geometry, including "spheres", "cubes", "triangles", and "surfaces", all in four dimensions. This is not only interesting in itself; it is also valuable for studying any phenomenon that requires four numbers for its description. To apply the four-dimensional metric to special relativity, the fourth variable must be related to the time. It cannot be time itself, because the units for distance and time are completely different, and, for any metric to describe a physical situation, the units in all terms must be the same. In fact, because the space coordinates have the units of distance, the fourth variable must also have the units of distance. This, along with the requirement that two different points at two different times are connected by the velocity of light, is enough to identify the appropriate time variable for the metric in the space-time four-dimensional continuum.

The difference between the space and time contributions to the metric is stark: the "time axis" is a continuum of imaginary numbers.[54] To emphasis the difference, the four-dimensional continuum thus defined is often called a $(3 + 1)$-dimensional continuum. Because of the different character of the time "dimension", the four-dimensional continuum is not completely analogous to

[54] Of course, there is nothing "imaginary" about imaginary numbers. They are just mathematical constructs such that their squares are negative. They were introduced into mathematics to deal with algebraic equations that had solutions whose squares were negative numbers.

ordinary three-dimensional geometry. In particular, the square of the metric is not always positive. It can be either positive, negative, or zero. This enhances rather than detracts from the utility of the four-dimensional space-time continuum.

In 1854, Riemann generalized the idea of distance to a form that was essential for the development of general relativity. In ordinary three-dimensional geometry, the metric is just the sum of the squares of the components of the distance between two points, when referred to Cartesian coordinates. Of course, in some other reference system, such as spherical polar coordinates, the metric looks more complicated, but in Euclidean geometry, a coordinate system (Cartesian) can always be found such that the distance is a simple sum of squares in which the coefficient of each term is unity. This is also true for higher-dimensional "spaces". In fact, the definition of a Euclidean space of N dimensions is that a metric can be found in which the infinitesimal distance is a simple sum of N squares of the infinitesimal components. Riemann's generalization was straightforward.[55] Just consider *all* quadratic forms for the distance, whether or not they can be reduced to a Cartesian formula of a simple sum of squares. If they could be so reduced, they defined a Euclidean geometry. If they could not, they defined a non-Euclidean geometry. Any space that could be defined by a general quadratic form is called Riemannian. Note that any Riemannian space, of any number of dimensions, forms a continuum. The mathematics of relativity is that of a four-dimensional Riemannian space with three spatial dimensions and one time dimension. For special relativity, this is a Euclidean four-space. In general relativity, the presence of gravitating masses makes the geometry non-Euclidean.

[55] This was an extension of Gauss' differential geometry of surfaces.

15

Time warps and bent space

The beauty of special relativity lies in the fact that it takes the laws of physics to be the same in every inertial coordinate system. It denies that there is any special inertial system that can be identified with an absolute space. But Einstein saw that this could not be the whole story. Shouldn't natural law be the same in *any* coordinate system, not just those moving at constant velocities with respect to each other? Why should inertial systems be unique? Why should laws of nature depend on the frame of measurement at all? In fact, if there were real differences between two coordinate systems, then that difference could be used to specify an absolute motion for one of them.

The essence of Einstein's work is that the laws of nature should be the same everywhere, no matter how they are described, and this is in direct opposition to how his theory has been interpreted by many non-scientists. The theory of relativity has been used to justify a general relativism in which there is no absolute truth. After all, if things are relative in the rigorous, mathematical world of physics, they must be relative everywhere. The misconception here is monumental. It is true that measurements of distance and time depend on the coordinate system in which they are made, but this is so because of deeper and more profound absolutes. First, the velocity of light in empty space is an inviolable absolute; second, the laws of nature are absolutely the same in all coordinate systems. Nothing could be less "relative". Again, this is an illustration of the old alchemists' concern that the uninitiated will misuse knowledge.

Einstein himself did not like the name "relativity". The name was actually attached to his theory by Max Planck and others. He preferred to call it "covariance", which stressed its universality and was in accord with his belief that the physical universe followed simple laws that were absolute in that they were true everywhere at every time.

Comparing special relativity with Newton's law for the force between two gravitating bodies exposes a problem. If it is applied to get the force between two celestial objects, for example, a simple question arises: what is the distance that should be used in the inverse square formula? In general, planets, moons, and stars are moving with respect to each other, and special relativity tells us that measured distances depend on the state of motion. Furthermore, in Newton's inverse square law, the time at which two gravitating bodies are acting on each other must be the same, but this is not true in special relativity. Even worse, the law of gravitation is often applied to bodies that are not in uniform motion. The paths of planets, comets, moons, and stars are far from straight lines.

None of this is accounted for in Newton's equation, so special relativity requires that the classical law of gravitation be modified. Its very existence tells us that special relativity is not enough and that something more general is needed.

Einstein's starting point was the fact that the motion of an object in a gravitational field did not depend on the mass of the object, but only on the strength of the field. This is just a restatement of Galileo's insistence that all bodies fall to Earth at the same rate, no matter what their mass.

The gravitational force acting on an object is proportional to the mass appearing in the law of gravitation. Newton's second law of motion says that the force is also proportional to the mass of the object that appears in the second law because of inertia. These are two different masses defined in two different ways and appearing in two different laws of nature. The first is called the gravitational mass while the second is called the inertial mass. Yet they are the same.

When the force as defined by the law of gravity is equated to the force as defined by the second law of motion, the two masses must cancel out if all bodies are to fall to the Earth with the same constant acceleration, as is observed experimentally. *That is, the gravitational mass must equal the inertial mass*. This is strange, but true.

The equality of the gravitational and inertial masses is called the principle of equivalence. It was implicitly taken to be a fact by Newton, and later verified to a high accuracy by experiments with balances on different kinds of materials.

Gravitation is a unique kind of force. First, it depends *only* on mass and is completely independent of the nature of matter. Metals, wood, water, gases, chemical compounds: all are subject to the same law of gravitational attraction. Electrical forces depend on matter being electrically charged, nuclear forces depend on the type of nucleons that are interacting, intermolecular forces are different for different kinds of molecules, and even springs exert forces that are different for different materials. Only gravity doesn't care what things are made of; for gravity, only the mass matters.

Second, there is no way to prevent the action of gravity. Appropriate shielding materials can stop the action of electrical or magnetic forces, but shielding is never possible for gravity.

Third, gravitational force has only one sign. It is always attractive and never repulsive. This is unlike any other force we know.

Maxwell's equations showed another major difference between electrical forces and Newtonian gravitation. Gravity was believed to be the ultimate example of 'action-at-a-distance,' in that it exerted a force that acted instantaneously throughout all space.

Electrodynamics, on the other hand, is local and is the most important example of a field theory. That is, electrical charges or currents create conditions in space that interact with charges. These conditions are said to constitute an electromagnetic field. The force on an electrical charge is the result of the action of the field in the space immediately adjacent to the charge. And the force does not act instantaneously throughout space; it propagates at a finite velocity.

Einstein saw the similarities between gravitation and acceleration, and illustrated this by his famous thought experiment of an elevator falling in the Earth's gravitational field. Inside the elevator, assuming the occupants cannot see outside of it, all physical measurements show that there is no gravitational field and all objects behave as if in an inertial system. If a passenger is holding something and lets it go, the object moves with the same velocity and in the same direction as the elevator itself, so it does not fall to the elevator floor. Inside the elevator, no experiment can be performed that would tell the observer the difference between being in a perfect inertial system and falling freely in a gravitational field. This is similar to the situation in Galileo's ship. If the ship is enclosed so that its occupants cannot see outside, there is no way they can tell if they are moving or not.

Also, if the elevator is in free space and some giant mechanism is pulling it in a particular direction at an ever increasing velocity, then when an object is dropped, the passengers see it fall to the elevator floor with an accelerating velocity. Unable to see outside the elevator, they conclude that a gravitational force exists, causing things to fall. The conclusion is that *the physics of material objects in a uniform gravitational field is just like the physics of material objects in an accelerated coordinate system.* This is another statement of the principle of equivalence and is often called the strong equivalence principle to distinguish it from the equality of inertial and gravitational mass, which is also known as the weak equivalence principle.

Einstein made it a foundation of his principle of general relativity, which said that the laws of physics had to be the same in *all possible coordinate systems*, not just inertial systems.

Note that if inertial mass and gravitational mass were different, accelerated motion *could* be distinguished from a gravitational field. For the elevator falling freely in a gravitational field, consider the case in which some object in the elevator is suspended in mid-air. The object has a gravitational mass and is therefore pulled down with the car. This is the mass that appears in the law for gravitational attraction; the greater the gravitational mass, the more it is attracted by gravity. The object also has an inertial mass, the mass in Newton's second law, which states that the greater the inertial mass, the greater the force needed to accelerate it by a given amount. Assume our object has an inertial mass of just the right magnitude to make the object move down in exact step with the elevator car. Now imagine that we double the inertial mass of the object. If its gravitational mass is more than the inertial mass, the object will move toward the floor of the elevator, because the object does not have enough inertial mass to oppose the force. Similarly, if the gravitational mass is less than the inertial mass, it will rise towards the ceiling of the car. But if the two masses are equal, the doubled mass will also move in step with the elevator.

Since the equality of inertial and gravitational mass has been verified to a very high accuracy, we conclude that a freely falling reference system cannot be distinguished from an inertial system. The conclusion is important enough to be repeated: *uniform gravitational fields and accelerated systems cannot be distinguished from each other by any physical experiment.*

There is an essential caveat here. The free fall of our elevator must be in a *uniform* gravitational field. Imagine that we have two balls alongside each other in the elevator, which is falling towards the center of the Earth. The field of the Earth is not uniform because it always acts in a direction towards its center. The balls then move along paths that get closer together as they fall, and this is an effect that can be seen inside the elevator. Now consider two balls that are at different heights in the elevator, one near the floor and the other near the ceiling. The gravitational field is stronger closer to the center of the Earth, so the acceleration will be greater for the ball near the floor than for the ball near the ceiling, and the distance between them will increase as the elevator falls. This means that a body falling freely in the gravitational field of the Earth will be stretched out in the vertical direction and compressed in the horizontal direction.

But if the variation of the gravitational field with distance is small enough, then, for a small region, the field is virtually homogeneous, and the equivalence of gravitation and acceleration is virtually exact. On the Earth's surface, for example, the acceleration of gravity is essentially the same for all heights of interest for daily life. In such a region, a coordinate system can be defined that is fully equivalent to an inertial system. For very large regions, however, it might not be possible to find such an inertial system. Near the Earth, for example, an inertial system at the North Pole can be defined for a region small relative to the size of the Earth. Just let the reference system be accelerated downward under the force of gravity. Similarly, an inertial system can be found for a small region near the South Pole. But no single acceleration can be found that can produce an inertial system that encompasses both regions. Therefore, a precise statement of the principle of equivalence is this: it is always possible to find a *local* inertial coordinate system that is equivalent to a *local* gravitational field. Just let the coordinate system have an acceleration equal to that of an object falling freely under the influence of the local field.

That is, all experiments give the same results in a local frame of reference in free fall as in an inertial frame of reference (far removed from all gravitating bodies). But gravity has not been defined away everywhere; no single inertial local frame can describe all of space, as illustrated by two elevators in free fall on opposite sides of the Earth.

But as far as the observers in one elevator are concerned, there is no gravity! In their local frame, they see all physics as if they were stationary, i.e. in an inertial frame. This means that any physical effect observed in an accelerating system will also occur in a gravitational field. Here is an important example. Mount a laser on the left wall of our car as it accelerates down a gravity field, and turn it on so that the light hits the right wall. In the free-fall frame, an observer sees the light move in a straight line to strike the opposite wall, because a freely falling system is equivalent to an inertial system. But what does the Earth-bound observer see? The laser must strike the same spot on the right-hand wall, but the elevator car has been moving downward at an ever increasing speed during the time it took the light to go from left to right. This means that, to the Earth observer, the light takes a downward-curving path to get from the left wall to the right wall. From our principle of equivalence, the same effect must exist in a gravitational field. We conclude that *a gravitational field bends light*. In his *Opticks* Newton had, in fact, speculated that masses would exert gravitational forces on light. The possibility occurred naturally to him, since he believed that light was corpuscular. But Einstein's conclusion was independent of the nature of light. It was based only on the principle of equivalence and is true whether light is a wave, a corpuscle, or, as in quantum mechanics, some combination of both.

Light rays are the ultimate rulers by which we make measurements in space. They give us the "straight lines" of physical geometry, so the fact that they bend in a gravitational field means that physical geometry is not Euclidean. The picturesque way of saying this is that gravitating masses bend space.

And there is a picturesque two-dimensional model that can help our understanding. Consider a large sheet of rubber with a heavy ball at its center. Because of its flexibility, the sheet will be depressed by the ball, and the rubber will form a funnel-like surface that is deepest near the ball and becomes less pronounced away from the ball until the sheet is almost flat at large distances. A line drawn from the center of the ball outwards defines a radius whose end describes a circle as it is rotated around the center. Now throw a marble onto the sheet and watch its motion. The marble will move towards the center of the depression. If the marble has a large enough velocity perpendicular to the radius, then the marble will describe a circular orbit around the center ball. If the velocity is very large, the marble will escape the influence of the bent sheet, and if it is too small, the marble will fall towards the ball. This is analogous to the relation between mass and space in general relativity. A massive object (the ball) distorts space (the rubber sheet) such that another mass (the marble) will fall towards it. If the mass being attracted has a very large velocity, it will move towards the attracting mass, but will escape. If the velocity is small, it will fall into the large mass. At intermediate velocities, the mass being attracted will go into orbit around the large attracting mass. What we call the gravitational field is the warping of space near an attracting mass. In our analogy, this is just the bending of the rubber sheet. Of course, the marble will also put a dent in the rubber, although a much smaller dent than that created by the large central mass. This is the analogue of the fact that attracted and attracting masses both have gravitational fields.

Notice that the falling elevator also produces a change in the frequency of light. To see this, consider a light wave traveling from the top to the bottom of the car. To an observer on the ground, the crests of the light waves are being pushed more closely together, because the elevator is moving towards him. The light frequency he observes is therefore higher than what he would see if the elevator were stationary. That is, there is a blue shift for the Earth-bound observer. Conversely, if the acceleration of the car is away from the observer, light suffers a red shift. Since an accelerating frame and a gravitational field are equivalent, we conclude that a gravitational field changes the frequency of light. The gravitational Doppler effect is in addition to the ordinary Doppler effect that arises from relative velocities. Note that this also demonstrates a time dilation effect of a gravitational field because light frequency is fully equivalent to a clock, each beat of the clock being one wavelength. Imagine an observer above the Earth or above some other massive body. Because of the gravitational red shift, he sees fewer wavelengths than normal and concludes that the clock in the gravitational field is running slow. This time dilation is in addition to that caused by velocity as described in special relativity.

Let's go further by considering a disc rotating relative to us and assume again that the laws of physics must be the same in all systems. For someone rotating with the disc, everything seems normal in that the disc is perfectly circular. In particular, if the observer rotating with the disc measures its circumference and diameter, he will get the answer π for their ratio, just as he would in an inertial system. But what does a stationary observer, sitting at the center of the disc, see? Each part of the circle is an infinitesimal line moving with a given velocity and is therefore decreased by an amount determined by the Lorentz contraction. The radius, however, is not contracted because it is always perpendicular to the circumference, which is the direction of motion. The ratio of the circumference to radius is therefore less than π. This result is impossible in Euclidean geometry. We therefore conclude that the geometry we see in a rotating system is not Euclidean. This is readily generalized to all accelerating systems, so the geometry of accelerating systems is not necessarily Euclidean. By geometry here, we mean *physical* geometry; the results of the actual measurements on actual objects. When these measurements are used to define geometric statements, such as the value of π or the sum of the angles of a triangle, they are not the same as those of Euclidean geometry. By the principle of equivalence, this must be as true for gravitational fields as for accelerated systems. Physical geometry and Euclidean geometry do not necessarily coincide. The properties of physical geometry must be determined by experiment.

Euclidean geometry is called the geometry of flat space, while non-Euclidean geometry is called the geometry of curved space. The nomenclature arises from the Euclidean plane, which is the simple example of a two-dimensional Euclidean space, and from the surface of a sphere, which is the prototypical example of a two-dimensional non-Euclidean space. Since gravity is equivalent to acceleration, we conclude that gravitational fields curve space.

Actually, it is space-time that is curved. Special relativity shows that space and time are not independent of each other. If the mathematics is worked out, it is found that the characteristics of space and time are indeed closely linked in relativity, and a four-dimensional geometry is required that includes both space and time. This sounds heavy but it is just a consequence of using ordinary logic.

The principle of equivalence led Einstein to his overarching conclusion that the laws of physics must be the same for *all* coordinate systems, whether they are inertial or not. Stationary, moving with constant velocity, accelerating, with or without gravitational fields: it doesn't matter. The laws of physics are identical when measured in *any* coordinate system.

It is now necessary to examine more carefully what we mean by "laws of nature" and what we mean by requiring them to be "the same in all coordinate systems". These terms are often used with several different meanings, each of which is normally clear from the context. Consider, for example, an object falling in the Earth's gravitational field after it has been dropped from a certain height. The distance it falls increases as the square of the time after it was dropped. This is called the law of falling bodies. Or consider a spring that is extended by some force pulling on it. The force on the spring is proportional to the length by which the spring is pulled out. This is called Hooke's law. Both of these are called laws of physics, but they have a specific rather than a general character. They each refer to particular instances of particular experiments. But Newton's second law of motion is much more general. It states that the force on any object equals its mass times its acceleration. And it applies to *any* force and *any* acceleration. It includes the force on a spring, the force of gravity, the forces between electrical charges, and even nuclear forces. It achieves this generality by being expressed in terms of rates of change.

The laws of falling bodies and of the extension of a spring cited above are given by algebraic expressions. That is, there is an algebraic formula that gives the height of a falling body in terms of the time, and another algebraic formula for the force on a spring in terms of its length. Newton's second law can include them both because it is a *differential* equation. That is, it describes rates of change of quantities rather than the quantities (time and distance) themselves. These equations are solved by using information specific to a particular physical circumstance. Specifying a particular force (gravity or spring constant) and particular initial conditions (position and velocity at some specific time) allows Newton's second law to be solved. In this way, descriptions for a large number of specific cases can be obtained. They include, for example, the orbits of the Moon and all other celestial bodies, the motion of gears, pulleys, and wheels, and the motion of electric charges. Newton's law has an enormous generality and applies to a huge number of physical phenomena. These are the kind of laws we mean when we say that they must be independent of which coordinate system we choose to express them.

This does *not* mean that the formulae for physical laws look precisely the same for all coordinate systems, even in classical Newtonian physics. The second law will look different if written in spherical polar coordinates,[56] rather than Cartesian coordinates. But there is a way of expressing the second law such that it indeed does have the same form no matter what Euclidean coordinate system is used, and that is to write it in vector form. Vectors are quantities that have direction, as well as magnitude. A vector is just like an arrow. The head of the arrow specifies direction, and its length specifies magnitude. Force is an excellent example because it is a quantity that has a magnitude acting in a particular direction. One reason vectors are so useful in science is that vector equations have the same form in any coordinate system.

If expressed in spherical coordinates, Newton's second law can be put in terms of the Cartesian coordinates by just using the geometric rules connecting the two systems. The points described in the two systems are the same: only the method of labeling has been changed. The physical content of Newton's law is unchanged. No matter what coordinate system we choose, Newton's law always gives the same physical result. Also, if two different coordinate systems are moving with constant velocity with respect to each other, Newton's law again gives the same physical results no matter which coordinate system is used. Furthermore, the equation is always the same, no matter what the coordinate system and no

[56] If a line is drawn in a plane, and a point on the line chosen as an origin, then any point in the plane can be located by measuring its distance from the origin and the angle between the reference line and a line joining the point to the origin. These two numbers are the polar coordinates for the point. Spherical polar coordinates are similar except that two angles and a distance are needed to identify a point in three dimensions.

matter what its velocity. It is always force equals mass times acceleration. The same is true for the dynamical equations of special relativity. They have the same form and give the same physical results for all inertial systems.

The equations are said to be *covariant*. This just means that all terms in Newton's equation change in the same way as we transform from one coordinate system to another, thereby keeping the vector form of the equation the same. The equations of both classical physics and special relativity are covariant for all inertial systems. Another way of describing covariance is to say that the fundamental laws of physics can always be described in terms that are independent of any particular coordinate system. Special relativity arose from the requirement that Maxwell's equations, as well as Newton's second law, be covariant with respect to translations among inertial systems.

Einstein believed that the fundamental laws should be the same with respect to *all* coordinate systems. Why should any one frame of measurement be better than any other? A physical event, such as the collision of two particles, should not depend on how the event is labeled, so physical laws should be the same in all coordinate systems, whether they are stationary, moving with constant velocity, accelerating, or rotating. That is, they should be universally covariant.

Of course, the trick was to find those laws of physics that were covariant for transformations so general that the only restriction on the coordinate systems was that a distance could be defined. It's not so hard for a given geometry. For a Euclidean space, such as a plane, or a non-Euclidean space, such as the surface of a sphere, covariant laws can be found by looking for transformations among coordinate systems within that space. But getting equations that are covariant for transformations among *any* set of coordinates, with either similar or different geometries, Euclidean or non-Euclidean, is a daunting task.

The mathematical machinery for such general coordinate systems had its origin in the work of Gauss and Riemann, and reached full expression in the work of Ricci and Levi-Civita.

These coordinate systems represent Riemannian spaces. This was described in the previous section, but they are so important that I will repeat and amplify some of that discussion here to connect it directly to Einstein's work on general relativity. Riemannian spaces are abstract, mathematical spaces in which it is possible to define a certain concept of distance. Consider a two-dimensional space; that is, a space in which it takes two numbers to label a point. Consider two such points. If this is our ordinary Euclidean plane, the distance between the two points is simply obtained from the Pythagorean theorem. For purposes of generalization, the two points are taken to be very close together, so the Pythagorean theorem states that the square of the differential distance between two points is the sum of the squares of its differential components. It is easy to imagine a space in which no such simple rule exists. The surface of a sphere is such a two-dimensional space. No coordinate system can be found on a sphere such that the square of the differential of distance is the simple sum of the squares of two components. There is a formula for the differential of distance in terms of the squares of components, and this formula is analogous to the Pythagorean theorem, but the squares in the formula are multiplied by functions of the coordinates. It is also easy to imagine a mathematical space in which the formula for the distance connecting two nearby points is *not* the sum of squares, even if the squares are modified by some function. In Riemannian spaces, distance is defined by a general sum of squares of differentials. If no such distance formula can be found, the space is non-Riemannian. Only Riemannian spaces have any meaning in physics, because, it must be possible to have a physical distance.

This is not as esoteric as it sounds, as is usually demonstrated by our simple example of comparing an ordinary spherical surface to an ordinary plane. On the plane, an infinitesimal distance is easily defined, and so is a finite length, and both of these are the same no matter where they are moved to on the plane. The surface of a sphere is different. If, for example, we try to move a segment of a great circle around the surface, it will fit only onto another great circle. It cannot be put onto a circle of a northern latitude without deforming it. The plane is the prime example of a Euclidean space, while the geometry on the surface of a sphere does not follow the rules of Euclid. For example, in the plane, the sum of the angles of a triangle is 180 degrees. But if we form a triangle on a sphere from three segments of a great circle, the angles between these curved segments sum to more than 180 degrees, a result quite different from that of Euclid. Also, while the shortest distance between two points in a plane is a straight line, on a spherical surface all distances between points are curved and the shortest distance between two points is an arc on a great circle. It is possible to construct a Cartesian coordinate system in a plane, but not on the surface of a sphere. But if we take an area on the sphere that is very small, then Euclidean geometry is approximately correct within that area, and the smaller the area, the better it approximates a plane. This is analogous to the fact that in a gravitational field, over a small enough region the field is constant, and a coordinate system can be defined that is inertial, but this cannot be done for large-scale non-homogeneous fields.

The spherical surface is a two-dimensional Riemanian space, while the plane is a two-dimensional Euclidean space. (Note that Euclidean space is just a special case of Riemannian spaces.) There is nothing conceptually weird about non-Euclidean, Riemannian spaces, but their mathematics is more complicated.

Einstein sought physical laws that were covariant with respect to coordinate transformations in the most general space that had physical meaning; namely, Riemannian space. More than any other scientific endeavor, the search for general relativity is the fruit of a mind that was totally free while being totally disciplined. Euclid's geometry is a statement of mathematics, not of physics. The experimental method demanded by physics would define geometry by the results of measurements with physical objects, not by abstract postulates that seemed reasonable, but may or may not be true in the real world. Einstein adopted the then radical position that the geometry of physics needed to be determined experimentally by real measurements, so it was not necessarily the same as the geometry of Euclid. He was totally free, not only of any preconceived notions of the meaning of space and time, but of any notion of the very geometry of physical space itself. At the same time, his commitment to observation and experiment was complete. Of course, he had to add another concept, which was philosophical in nature; namely, that no particular place or time in the universe was special compared to any other place or time. The fundamental laws of nature had to be the same everywhere, no matter which coordinate system was used to describe them. In relative motion, acceleration, or rotations, no matter what, the laws had to be the same. While the beginnings of this concept might be regarded as philosophical, the development of science has shown it to be an experimental fact.

The first concept to examine in any geometry is the meaning of distance. For the space-time continuum this means finding the metric for a four-dimensional Riemannian space that gives the shortest possible path between two points. This is a familiar problem in ordinary calculus, where it is often necessary to find smallest values of a variable or of a function. Just write down the expression for the distance between two points and work the known mathematical machinery for finding its lowest value. Of course, the process is more complicated for the space-time continuum than in ordinary space, because the space we are dealing with is four dimensional and not necessarily Euclidean. And it looks complicated when written out, because there have to be labels that describe each dimension and the coordinate transformation properties for each dimension. But the process is straightforward and the result is a set of equations that can be solved to give the shortest distance between two points for any Riemannian space. Applying this to three-dimensional Euclidean space gives the equation for a straight line, while using the metric for the surface of a sphere gives the shortest distance as an arc on a great circle. The shortest distance between two points is called a geodesic. In general, the process gives the geodesic for any Riemannian space. This is a purely geometric concept arising from purely mathematical properties.

Einstein converted this into physics by taking the geodesic in the space-time continuum to define a generalized law of inertia, which states that the *motion of any object not subject to any forces is along a geodesic*. There is a multiple rationalization for this assumption. First, the law of inertia for classical mechanics does state that in the absence of forces a body moves along a geodesic, which in that case is just a straight line. Second, for inertial systems, constant velocities can always be made to vanish by transforming to the appropriate coordinate system. The generalization to general motion is that, again, motion in the absence of forces is a geodesic, and, for accelerating systems, it is always possible to find a coordinate system in which the acceleration from gravity is zero, if no forces are present. In this context, gravity is not considered a real force, because it can be transformed away. This is not true for other forces. The force between electric charges, for example, cannot be reduced to zero simply by going to another coordinate system, because electrical forces are non-zero whether looked at from a stationary or an accelerating coordinate system. In general relativity, "force free" means "under the influence of gravity only".

Einstein was then able to build on the idea of the geodesic to develop equations of motion for a mass, analogous to Newton's second law of motion. To do this, it was necessary to use the fact that for very weak gravitational fields and small velocities, the law of gravity, and the law of motion, must reduce to the Newtonian laws. For high velocities, in the absence of gravity, the equations reduce to special relativity.

The covariance requirement and the principle of equivalence then gave equations of motion whose form was the same for all physically possible transformations regardless of their state of motion, not just for those with constant relative velocities.

General relativity modified our concepts of space and time beyond the already revolutionary results of the special theory. Bodies move along geodesics in a curved space-time, and the equations show that the curvature of the four-dimensional continuum is completely fixed by the distribution of masses. The masses curve space-time, and massive bodies move along the resulting curvature. Here is the resolution of the mysterious fact that inertial and gravitational masses are equal. General relativity asserts that there is no mystery, because bodies move along geodesics without requiring any force to act on them. In the absence of matter, these geodesics are just straight lines, and we have the classical law of inertia. When matter is present, however, it induces a space curvature, and the geodesic is no longer a straight Euclidean line, but a curve along which objects move. The orbit of the Moon around the Earth is simply an example of an object moving along a geodesic *in the absence of forces.*

Note that general relativity has gotten rid of the idea of action-at-a-distance and replaced it with a field theory in which the gravitational force at any point is the result of the curvature of space at that point.

What an achievement! From a fundamental belief that natural laws must be the same everywhere, Einstein reconstructed the entire foundation of physics, using only a few experimentally observed facts. It was the most far-reaching revolution in science since Newton and was matched only by the development of quantum mechanics.

Most physicists quickly recognized special relativity as valid. The holdouts were those who simply could not accept something new if it seemed strange, those who learned of the theory through hearsay rather than by reading Einstein's work, or those who were blinded by ideological considerations. General relativity was mathematically more difficult, and the theory of differential geometry in Riemannian spaces using tensor analysis was not a part of physics curricula at the time. Einstein was fortunate in that the definitive paper by Ricci and Levi-Civita was published in 1899, so he had tensor analysis at hand. This mathematics is admittedly more complex than that normally taught to scientists but certainly not beyond the capabilities of good graduate students. Again, it was the strangeness of the results that was so off-putting.[57]

[57] However, general relativity was not widely studied by many physicists for about a decade. Eddington was among the very first to do so and wrote an

To compound the strangeness, it is not merely three-dimensional space that is curved; it is the four-dimensional space-time continuum. How can we, who have grown up with our intuitive concepts of time and space, accept such a thing? Most of us do not even know what a "curved space-time continuum" really means. Some of this weirdness is just the result of unfamiliar language. In particular, the phrase "four-dimensional space-time continuum" sounds as if it is full of mystery, but it is merely scientific shorthand for something quite simple. An event in physics, such as a collision or the end of a rock's free fall, takes place in space at a particular time, so it requires four numbers to specify it; three for its spatial position and one for the time at which it occurred. The event is therefore said to be four-dimensional. The fourth dimension of relativity is not at all like another spatial dimension, perpendicular to the other three. That would be nonsense.[58] But the mathematics of relativity brings the time and the spatial dimensions together, shows their intimate relationship, and treats them as parts of a single continuum of "points", each being specified by four numbers. The results of relativity are strange enough without the unnecessary confusion of misunderstood language.

outstanding book called '*The Mathematical Theory of Relativity*.' He was certain of his abilities. When asked if it was correct that only three people understood relativity, he responded by saying he was trying to think of the third one. This early commitment to the theory might have led him to overestimate the accuracy of his observations on the deflection of starlight by the Sun. It would not be the only time that a scientist would help "wish" his results to an expected answer.

[58] Physicists have proposed theories with more than four dimensions. In 1921 T. Kaluza tried to unify gravitation and electrodynamics using a five-dimensional theory. But he was careful to construct it in such a way that the fifth dimension had no effect on the observed dimensionality of space-time. In modern string theories, extra dimensions are "curled up", so they have no effect on ordinary space. Within the realm described by general relativity, a fourth spatial dimension is still nonsense.

It helps to keep in mind that the meaning of space and time in physics is nothing more than the results of measurements with rulers and clocks. Physical space is not the same as mathematical space. Physical space is just the result of physical measurements. A curved space just means that if measurements are made with measuring rods, the results do not obey Euclidean geometry. This is strange to us only because our entire history has been spent with systems that are small with respect to cosmic distances, and with velocities much less than that of light. Two-dimensional creatures who have always lived on the surface of a sphere would not find their curved geometry strange at all and would recognize that Euclidean geometry holds only when very small regions of the surface are considered.

This is not the end of the story. Einstein recognized immediately that the universality of physical laws he was looking for needed the theory of gravitation, so well described by general relativity, to be combined with the theory of electrodynamics, which was still well represented by Maxwell's equations. After working out general relativity, he spent the rest of his life looking for such a theory. He was not successful. The modern search for a unified field theory attempts to unite quantum mechanics and relativity and to bring nuclear forces, as well as gravitation and electrodynamics, into one all-encompassing scheme. The program is not yet complete.

16

It stands alone

Classical mechanics and relativity are unlike many other physical theories in that they are not based on any ideas of what the constituents of the world are like or what they are made of. They do not depend on whether or not matter is made of atoms, or on whether light consists of particles or waves, or on the kind of forces acting between atoms, or on any other information about the nature of matter and energy. The only other theory with such simplicity is thermodynamics, which is also independent of material constitution and of time and place. In fact, some scientists were so impressed by its universality that an entire school of scientific thought rose up in the nineteenth century, called Energetics, that tried to base all science on thermodynamics without any reference to atoms. But mechanics is different in several respects. First, it includes the crucial concept of time; second, it connects widely different regions of space, whereas thermodynamics is essentially local; and third, it encompasses a description of dynamics. In many ways, thermodynamics is a flawless logical construct that exposes the relations among physical quantities. But if any significant numerical results that relate to real physical systems are to be obtained, the underlying theory of statistical mechanics that relates thermodynamic quantities to the nature of atoms and molecules is required. Classical and relativistic mechanics, on the other hand, gave many new results, and continues to give new results, without reference to the nature of matter.

Elasticity theory is based on assumptions about the forces acting between elementary parts of solid bodies; theories of chemistry are based on the nature of the fundamental interactions among atoms; electrodynamic theory is an expression of the nature of electrical forces; hydrodynamics arises from the properties of fluids. And quantum mechanics, the other great scientific achievement of the twentieth century is, from top to bottom, a study of the ultimate composition of matter and the properties of its constituents.

But relativity is quite different than classical mechanics. It is nothing more than the study of what it means to make scientific measurements of distance and of time. It grew out of an analysis of the actual physical operations of making a measurement of distance with a real physical ruler and of measuring time with an actual physical clock. It is true that many of the physical results depend on specific laws that do arise from applying relativity to theories of the nature of matter and radiation, but this is not the essential point. Relativity is, fundamentally, nothing more than a theory of the simplest of all experiments: the measurement of macroscopic distances and times.

The foundation of relativity is incredibly basic and incredibly simple, and yet it took men of genius to discover it and to develop it. It took genius to accept the results of experiments, even when they ran counter to conventional wisdom and established beliefs, and it took genius to follow the logic based on these experiments rigorously and without deviation no matter how strange the consequences.

Human intelligence did not evolve for the purpose of discovering the truths of nature, but in response to the requirements of survival. Manipulation of matter on a human scale was important; time intervals of minutes, hours, and months were important; speeds of rivers, running animals, and visible projectiles were important. Human-scale times, distances, and speeds were the experiences that fed intelligence and created the physical intuition that allowed human beings to survive and prosper. This intuition even predated the emergence of human consciousness, because animals also had to deal with a world governed by the same scale. So when scientific studies showed that these concepts had to be changed in ways that made them completely different from common experience, the mind strongly resisted the new concepts. They were at variance with ideas that had served people well throughout their existence, and only genius could break through the conditioning of millennia.

Field theories are intrinsically beautiful because they start with a few assumptions from which all else follows by the application of elegant differential equations. Relativity is the most elegant example of a field theory.

All who have studied the theory of relativity have found it hugely attractive and called it beautiful. It is so compelling and carries such an aura of truth that its verification by observation and experiment is indeed fortunate. If relativity were not true, then there would be something wrong with our perception of the world, because it feels as if it surely *should* be true. Of course, experimental fact is the ultimate arbiter, but what a loss it would be if relativity were shown to be false! The basic concepts of relativity have the feeling of always being right.

The elegance of relativity theory starts with the simplicity of its assumptions. Let's restate them here and stress that they apply to general relativity, of which special relativity is a limiting case. The first is that the speed of light in a vacuum is the same in every local inertial system. The second is that the laws of nature are the same in every coordinate system. The first assumption is the theory's experimental content and the second is its philosophical foundation. It's hard to argue with these statements. Every measurement ever made supports them and no violation has ever been found.

There is a profound difference in the nature of the two assumptions even though they are both empirically based, and I have tried to recognize the difference by calling one of them "philosophical". The constancy of the velocity of light is a straightforward fact and readily verified by specific measurements.

The second postulate is called the "principle of covariance". It is not immediately established by any single type of experiment. To the modern scientific mind it is a natural assumption because it seems ridiculous to think that the laws of nature should depend on which coordinate system is used to describe them. But this was not always so, and science had to evolve for centuries before such an idea could be accepted.

Aristotle and Ptolemy thought that the Earth gave a special point of view for all natural phenomena, celestial as well as terrestrial. This was first adopted on scientific grounds but was later given the force of religious dogma. Copernicus revived Aristarchus' ancient idea of a heliocentric solar system, and this was confirmed by Galileo's observations, so the center of the world moved to the Sun, but there was still a special set of coordinates for describing nature. It was not until Faraday and Maxwell forced Einstein to carefully analyze the measurement of distance and time that it became clear that *no* coordinate system was special. The history leading to covariance is the story of a continual march away from anthropomorphism: from man being the center of all things, to every point in the universe being equally valid for describing nature.

Only logic is used to work out the consequences of relativity. No other data or assumptions are needed. The mathematical structure is so general that it does not even refer to any particular geometry or set of coordinates, but is valid for all of them. Furthermore, the theory is fully deterministic, in the sense that it ties together causes and effects in a tight temporal chain, and all quantities can be defined and computed to any degree of precision required. Also it is a field theory, which means that any event at a particular point in place and time is determined only by the conditions of the space very near to that place and that time. There is no action-at-a-distance, an idea that was always troublesome.

A measurement of time or distance is well defined in relativity theory and always gives a unique answer. The limitations of the theory arise from the fact that these are macroscopic measurements that ignore the atomic constitution of matter. It is assumed that measurements of distance or of time can be made to any degree of accuracy desired. But we know that such accuracy is impossible, even in principle, because experimental devices are made of atoms. We know, from quantum mechanics, that any measurement at atomic distances disturbs the system, so that precise measurements are impossible. The process of measurement transfers energy to the atom that kicks it around, so position cannot be measured precisely, and distance is indeterminate. This is true for all measurements. At the atomic level, measurements are subject to Heisenberg's uncertainty principle, which states that simultaneous measurements of physical quantities can be determined only within certain limits. This is not because devices cannot be made accurately enough; it is an inherent property of nature.

General relativity is a macroscopic theory, and it breaks down for the description of nature at atomic scales. Normally, this is not a problem because gravitational forces are very weak compared to other forces. Intermolecular, interatomic, and internuclear forces are so much stronger that gravity can be completely ignored when one studies molecules, atoms, or fundamental particles. And gravitational effects normally involve such large distances and massive objects that interactions at the atomic and nuclear scale are irrelevant. Usually, then, relativity, which is a theory of measurement of macroscopic objects, and quantum mechanics, which is a theory of measurement of microscopic objects, are well separated and work well within their respective domains. But there are important exceptions, and these are *very* important exceptions. When a region of space is very small, so that it is of the order of atomic dimensions, and its mass is so high that its density approaches infinity, then a theory that simultaneously describes quantum-sized distances and very large gravitational forces is needed. The interior of a black hole and the Big Bang origin of the universe are two such instances, so there are astrophysical and cosmological phenomena that cannot be understood unless relativity and quantum theory can be combined into a single, unified whole. The search for such a theory has been going on for decades without success. The fundamental difficulty is that the two theories are based on radically different concepts of that most important issue, the meaning of a measurement.

Yet relativity is unparalleled. The stark simplicity of its assumptions, its striking mathematical structure, its universality, its deterministic field theory character, and its total success in its own domain make it unlike any other scientific theory.

The search for its unification with quantum mechanics is just another expression of Einstein's belief in the unity of nature and his confidence that the ultimate answers could be found by human reason.

17

This too is true

It was the prediction of the bending of light by the gravitational field of a star that captured the public imagination and catapulted Einstein to celebrity status when it was announced that Eddington had confirmed his results. But there was serious doubt that Eddington's observations actually did agree with general relativity. The images of starlight bending around the Sun during the solar eclipse were hard to measure and hard to interpret, since the displacements of the images of stars from their normal positions were so small. Also, Eddington omitted some of the data when making his analysis, an act that always makes scientists suspicious.

Actually there were two expeditions sent to observe the May eclipse. The more famous one, led by Eddington, was to the Island of Principe off West Africa, and it was the results of this expedition that made such an international sensation. The other was led by Crommelin and Davidson, who went to Sobral in Brazil, where they made observations of the same eclipse. The purpose of both expeditions was to look at stars as their light grazed the Sun, which was only possible during a total eclipse. Einstein's and Newton's theories of gravitation both predicted that light would be bent by the Sun's gravity, but Einstein's prediction was twice the size of that from Newton's theory.[59] A comparison of stellar positions

[59] Note that if light is a corpuscle, as assumed by Newton, it is not necessary to know its mass to determine its orbit around a large body because of the equality of gravitational and inertial mass. It is only necessary to know the mass of the large body. Thus the gravitational effect on light is easily calculated from Newton's theory.

when observed near the Sun with observations when they were far from the Sun should show the magnitude of the deflection, thereby showing whether Einstein or Newton was correct. Eddington reported that Einstein was right, and his enthusiasm grew with time. At first he was doubtful that the plates were definitive, but later called the observations the greatest moment of his life. The Sobral observers made no positive claims and concluded that they could not verify the general relativity results.

Ultimately, the Einstein prediction was indeed verified by a variety of methods, including radio astronomy and observations of gravitational lensing, but Eddington's observations were deemed too inaccurate for a definitive conclusion. The scientific community, therefore, did not unequivocally accept Eddington's pronouncements with the same wild enthusiasm shown by the media and the public.

Later observations of starlight being bent around the Sun in eclipse after eclipse verified Einstein's theory beyond any possible doubt. But the Eddington and Crommelin expeditions did not provide that proof.

Eddington's problem was that he already believed in general relativity and he fully expected to find it confirmed. He learned about the general theory in 1915 and immediately became convinced that it was correct. In fact, he was so sure of it that he felt no need to go on the Principe expedition at all! The theory had an irresistible fascination for him. He was the first to study it completely and thoroughly and was a major force in bringing it into the mainstream of physics. His book *The Mathematical Theory of Relativity*, published in 1923, was based on his Cambridge lectures and was called the "finest presentation of the subject in any language" by Einstein himself. It appealed to a mystic streak in him: that urge felt by many scientists to find the overarching scheme that dominates the universe. Eddington succumbed to this urge in his later years by trying to work out an overall synthesis of the world based on relativity and quantum mechanics. This was given expression in his book *Fundamental Theory*, which relied heavily on the analysis of numerical constants. To many, including this author, it was barely intelligible and seemed to slip into mystic numerology.

Eddington was born in 1882 and studied physics and mathematics at Manchester and Cambridge. He was able to indulge his true calling in 1906, when he was appointed to a position at the Royal Observatory at Greenwich, becoming one of the greatest astrophysicists of the twentieth century. He was the first to realize that the temperatures in stellar interiors were extremely high and that all atoms in stars were completely ionized, and he discovered the mass–luminosity relation[60]. Also, he was the first to recognize that the energy produced in stars came from nuclear reactions, which he correctly identified as hydrogen fusion. He was an excellent writer, and, in addition to his research, he wrote a number of outstanding science popularizations for a general audience.

No matter how inexorable the logic from philosophically satisfying principles to the final results, and no matter how internally consistent, a theory can be accepted only if it agrees with observation. The bending of light in a gravitational field was the general public's contact with that agreement.

But the first and simplest test is that ordinary Newtonian mechanics must be found to be correct to a high degree of accuracy for most cases. The classical theory has been so right for so many phenomena that any theory that totally rejects it cannot be correct. Indeed general relativity passes this first test. If gravitational forces are small, general relativity reduces to special relativity, and if, in addition, the velocities of moving objects are small relative to that of light, we recover Newtonian physics. Without this circumstance relativity cannot be correct, because Newtonian physics worked well for practically all phenomena for over two-and-a-half centuries and ran into trouble only when velocities were high and gravitational fields were very strong.

[60] The greater the mass of a star, the greater its luminosity. The dependence is quite strong. Doubling the mass of a star increases its luminosity more than eleven times.

But this is not completely convincing, because relativity was obtained by requiring it to reduce to classical physics as a first approximation, so this requirement is built in. Nevertheless, it is important that it is *possible* for general relativity to reduce to Newtonian mechanics. If this could not be done, the theory would have to be rejected.

But predictions of new effects must be tested and verified. The equations Einstein obtained for general relativity did verify the conclusions obtained from the elevator thought experiments on the equivalence of gravity and acceleration and went much further to give a complete relativistic theory of physics. Here is a list of some results that can be tested by experiment:

1. Gravitational fields deflect the path of light rays.
2. Gravitational fields shift the frequencies of light. (gravitational red shift). Since light frequency is a measure of time, this is the same as saying that gravitational fields change the measurement of time.
3. The orbits of celestial bodies are slightly changed by general relativity effects.

If these effects can be measured with sufficient accuracy, and if they are found to be in numerical agreement with the predictions, then general relativity is confirmed.[61]

Gravity is a weak force. The Earth has a mass of 6.6 sextillion tons (a sextillion is a thousand billion billion, where a billion is defined by the American convention of being one thousand million.) and yet we can totally escape its pull with a velocity of a little more than 11 kilometers per second (about 7 miles per second), which is readily attained by modern rockets. Electrical forces are so much stronger that gravity can be totally neglected in practically all electrodynamics applications. This means that the effects of general relativity are usually very small and very difficult to detect. Modern instrumentation has made it possible

[61] Note that no matter how many experiments confirm a theory, it cannot be proven to absolutely correct. It only takes one contrary experiment to show that a theory is wrong. Any good theory must be falsifiable.

to measure these effects with extraordinary precision, and every observation so far has verified the predictions of general relativity.

The deflection of light by gravitational fields has become a standard tool in observational astronomy because it accounts for apparent anomalies in the positions and motions of celestial objects and provides information about them that would not otherwise be available.

The bending of light by massive objects is called gravitational lensing, because it is fully analogous to the bending of light by a refractive medium. It is just as if mass induces a variable refractive index in space, thereby changing a light beam's velocity and causing it to bend. The proper description in terms of general relativity is that mass creates a curvature in space-time and that light follows the resulting geodesic.

A spectacular example of gravitational lensing was observed in 1976 when two identical quasars about six seconds of arc apart were seen to have identical spectra and identical red shifts. This was such a highly improbable coincidence that the two images had to be from the same object. The light from a quasar[62] passed close to a massive galaxy whose gravitational field bent the light around two of its sides, thereby sending two images to the observer. Multiple images of many quasars and of galaxies have been found since then, and gravitational lensing has become a standard tool for many kinds of astronomical studies.

The bending of light, the change in its frequency, and time dilation in the presence of gravity were seen to be new effects that must exist if the full principle of equivalence is correct. The full theory of general relativity shows that the effects we deduced from examining an elevator in free fall all come right out of the equations.

[62] Quasar is shorthand for "quasi-stellar object". Such objects are enormously distant from us and emit incredible amounts of energy. They are small enough to look like stars but generate more energy than many galaxies. The lensing effect giving rise to a double image was observed in 1979 at Kttt Peak Observatory by Walsh, Carswell, and Weymann.

The bending of light in a gravitational field leads to an effect that is observable in our solar system. If a radar signal is bounced off the Moon or a planet and reflected back to Earth, the time for the round trip will be different if the beam is near a massive body. A nearby mass curves space, and the beam takes longer to make the round trip because it takes a curved path. So if we bounce a radar beam off a planet, the time for the round trip will be longer, the closer the path is to the Sun. This effect is actually observed.[63]

The simple thought experiment with elevators is remarkably fruitful. The red shift predicted by the mathematical theory was measured in 1960.[64] The gravitational red shift should also shift the spectral lines of the Sun and this has also been observed.

A direct verification of the time dilation accompanying the gravitational red shift (gravitational Doppler effect) was made in 1971 by measuring the times on clocks flying around the world.[65]

An interesting application of general relativity is in the Global Positioning System. Use of GPS requires that the time of light travel be accurate within 30 nanoseconds to get positions accurate within 10 meters. A calculation of the effect of gravitational fields on the velocity of light is needed to attain such accuracy.

The general theory is now widely applied in astronomy, where its optical consequences must be accounted for in all stellar observations, and in astrophysics, where it is essential whenever large gravitational fields occur. These include the fascinating compact objects such as white dwarfs, neutron stars, and black holes.

[63] Shapiro first proposed this in 1964.

[64] Pound, Rebka, and Snyder used the Mossbauer effect to detect a difference in frequency of 14.4 keV gamma rays from Fe^{57} sources at a height difference of 22.6 meter (the top and bottom of Harvard Tower). and found a shift of 5.1×10^{-15}. The theoretical value is 4.9×10^{-15} (2% error).

[65] In October 1971, J. C. Hafele and Richard Keating flew four cesium atomic clocks twice around the world, first east and then west. Special relativity slowed the eastbound clock (it lost time) and speeded up the westbound clock. General relativity caused a time gain for both directions. Adding the results for the two directions gave an answer in agreement with general relativity theory.

General relativity predicts that gravitational waves must exist and that their velocity is that of light. They are transverse waves and can exist in two states of polarization. In these respects they are similar to electromagnetic waves. Gravity is the result of a distortion of space-time by matter, so if matter moves, there must be changes in the space-time continuum. Imagine, for example, that a large chunk of matter vanishes. The surrounding space no longer has any reason to be curved, so it goes flat. Think of the analogy of the rubber sheet with a massive ball in its center and now remove the ball. The rubber springs back into a flat sheet. Of course, it takes some time for the sheet to straighten out, and during that time, a wave travels through the sheet. Also, if the ball is not taken away, but just moved from one place to another, the deformation of the sheet will change, and this change will propagate with a speed controlled by the elastic properties of the rubber sheet. In the same way, a wave will spread through space when a mass is moved. Since gravity is just distorted space, this spreading is a gravity wave. Laboratory attempts to detect these waves have been unsuccessful so far[66] but there is strong evidence for their existence from observations on double stars. As two stars rotate around each other, there should be a decrease in their period of rotation because the gravitational waves radiate energy away from the double star system. Taylor and Hulse saw this in a neutron double star, whose periods are decreasing at a rate in good agreement with the results of general relativity.[67]

[66] A facility for detecting gravity waves exists and is under continual improvement. LIGO (Laser Interferometer Gravitational-Wave Observatory) was started by Kip Thorne and R, Drever from CalTech and R. Weiss from MIT in 1992. It is getting continual support from NSF and there is an expectation that the enormous difficulties of getting enough precision can be overcome.

[67] In 1993 Taylor and Hulse received the Nobel prize for their observation in 1974, and correct interpretation, of this phenomenon in a binary pulsar.

If gravitational waves could be detected, a unique, exciting observational tool would be added to astronomy. Imagine being able to see the collapse of matter into a black hole, or to follow the rotation of two black holes around each other, or their coalescence into a single black hole. With enough sensitivity, it would be possible to track rapid astronomical processes from the movements and changes of massive objects.

The bending of light in a gravitational field was not the first experimental verification of general relativity. In his 1916 paper, Einstein pointed out that there is an important non-optical effect: general relativity predicts that planetary orbits are not quite the perfect ellipses given by Newton's theory of gravitation.

Understanding planetary motion was the great outstanding achievement of Newton's theory. The derivation of Kepler's laws and the use of the inverse square law of gravitational attraction to calculate the orbits of the celestial bodies in the solar system was the most stunning success of Newton's work. The orbit of a planet around the Sun is a perfect ellipse if the only force acting is that between the Sun and the planet. Other celestial bodies, however, also exert forces on the planet, so its orbit is not quite an ellipse. But these other forces can be included in the calculations, and when this is done it is found that the data are in extremely close agreement with the calculated orbits.

Except for Mercury.

No matter how carefully the effects of all other planets were taken into account, the orbit of Mercury did not quite match the calculations from Newton's theory. There was a small rotation of the perihelion that could not be accounted for. The perihelion is the point on the elliptical orbit that is closest to the Sun, and if no other celestial bodies existed except Mercury and the Sun, it would always be constant. Other planets, however, perturb the orbit, so the perihelion moves, describing a rotation around the Sun. The disagreement of this motion, called the advance of the perihelion, with the detailed Newtonian calculations was first found and studied by the great French astronomer U. J. J. Le Verrier in 1855, who found the discrepancy to be 43 seconds of arc per century.[68] He postulated that an undiscovered planet existed that perturbed Mercury's orbit. He called this planet Vulcan and thought that its orbit was closer to the Sun than Mercury's, at an appropriate distance with an appropriate mass, to account for the observation. He had successfully predicted the existence and position of Neptune from deviations in the orbit of Uranus, which was found by Johann G. Galle at Berlin in 1846. His speculation was therefore taken very seriously. But the presumed orbit of Vulcan was very close to the Sun and therefore difficult, if not impossible, to observe. In fact, no such planet was ever found.

The matter was resolved on the last page of Einstein's 1916 paper on general relativity in which the relativistic value of Mercury's perihelion advance was calculated to be in agreement with observation.

Perihelion advance is observable only for large gravitational forces, which is why it was seen only for the planet closest to the Sun. Actually it is just possible to see it for Venus and Earth, and again the results of the theory agree with observation (about 8 seconds of are per century for Venus and about 5 seconds of are per century for Earth.)

So far, every test that has been made confirms the results of general relativity, and no experiments or observations have been found that contradict it.

[68] This is a small effect. It would take over three million years for the advancing perihelion to make one complete rotation around the Sun.

18

Crunch

Gravity is inexorable. Its force is exerted by every bit of mass, and it works on every bit of mass. Nothing can shield it, and its range extends to infinity. Any aggregate of matter, whether nuclei, atoms molecules, dust particles, or stars, will ultimately coalesce into a contiguous mass unless there is some force strong enough to resist the mutual gravitational attraction of its parts. It gives rise to fantastic objects and to incredibly violent processes and is indeed responsible for the very structure of the universe. Supernovae, white dwarfs, neutron stars, and black holes, the formation of stars and their evolution, even the existence of carbon, on which life is based: all come about because gravity works everywhere and always.

Gravity is one of the four fundamental forces of nature. The others are the electromagnetic, the strong nuclear, and the weak nuclear force. But gravity is so much weaker than the others that it seems to be in a class by itself. The electromagnetic force is incomprehensibly stronger than gravity. In fact, it would take the gravitational mass of more than 27,000 Suns, to equal the attractive electric field of one gram of electrons.

The electromagnetic force controls many of the phenomena of our daily lives. It is the origin of the chemical bonds in the enormous variety of all the materials we use. Metals, ceramics, polymers, drugs, foods, and our very bodies are all structured and function through the action of the electromagnetic force. Its variation with distance is just like that of gravity: the force between two charges falls off as the square of the distance between them. But there is a major difference. There are two kinds of charges, positive and negative, so there are two kinds of electric forces, attractive and repulsive, whereas gravity is always attractive. In practice, therefore, the electromagnetic force does not have a long reach, because there are an equal number of positive and negative charges that attract each other. This is wonderful for forming atoms and molecules, and the chemistry and physics of everyday matter is controlled by the electromagnetic forces. These are so strong over short distances that gravity can be completely neglected for ordinary materials. But the long-range influence is weak, because the opposite charges effectively neutralize each other and the net forces they exert fall off very rapidly with distance.

Nuclear forces are also very large over short distances: they are strong enough to hold atomic nuclei together in spite of the immense electrical repulsions between protons. But their range is quite limited, much more so than for electromagnetic forces; the nuclear forces are essentially zero for distances beyond the diameter of an atomic nucleus.

Gravity is the only force that spans the universe, the only force that is always present and active and always attractive. Newton realized that gravitational force draws all matter together and that everything should have collapsed into a single mass long ago. But, he said, this is true only if the universe is finite. For an infinite universe, there is no central point for mutual attraction, and the attraction on any mass is the same in all directions, so collapse does not occur. While he thought that this resolved a paradox, Newton recognized the possibility of gravitational collapse and its close relation to cosmology.

Others who knew of Newton's theory of gravity speculated on its extreme results, and in 1783 John Michell concluded that there could be objects so massive, and therefore with an escape velocity so high, that even light could not escape from them. Michell is remembered primarily as a geologist and is credited as a founder of the science of seismology, but he was also an astronomer who had considered the existence of binary stars.

Michell's idea was later taken up and enlarged by Laplace, who stated that the largest objects in the universe might not be visible. But the competing theory of light being a wave won out as interference phenomena were studied, and, since waves were not subject to gravity, the idea that a large mass could prevent light from escaping was dropped.

The Michell conjecture was the first proposal for the existence of black holes (a name invented in 1964 by John Wheeler).

Any aggregate of matter will continue to condense until gravity is opposed by an internal pressure tending to drive its particles apart. In a gas, this pressure is supplied by the motion of the gas molecules and is larger the higher the temperatures and densities. In liquids and solids, the internal pressure arises from the repulsive forces of the electrons in the outer shells of atoms.

Gravity is a weak force but for large objects it is very important, and we know how it varies with mass. An object must be quite large before its self-gravity can overcome the internal pressure from the interatomic electrical repulsions. In fact for masses smaller than about one tenth that of the Sun (actually about eight percent), these repulsions work fine, and planets are all objects with masses less than this limit.

Consider a hydrogen cloud with the individual atoms far enough apart that the only force acting among them is gravity. The history of this cloud depends on how much mass it contains. At first, the atoms will attract each other no matter what the mass, but sooner or later repulsive forces come into play that keep it from condensing further. If the mass is relatively small, the repulsive forces among the atoms are enough to balance the attraction of gravity, and an equilibrium is reached similar to that for the Earth, whose size is fixed by the balance between electric repulsion and gravitational attraction.

Two conservation laws must be considered when following the history of a coalescing cloud of gas. The first is that the total energy of any physical system is a constant. When the atoms are far apart, they have a potential energy arising from their mutual gravitational attraction. As they come closer together, this potential energy decreases, but the atoms acquire kinetic energy as they come together with increasing speeds. When the inrush of atoms is stopped by the repulsive forces among the atoms, the kinetic energy manifests itself as heat, and the temperature of the aggregate goes up.

The second conservation law is that of angular momentum. If the cloud is rotating around some point, then the angular momentum stays the same as the cloud condenses. Since the angular momentum is conserved, the angular velocity must increase when the cloud gets smaller. The commonly used example to illustrate this is of an ice skater spinning on her toes with her arms outstretched. If she quickly draws her arms to her side, she will immediately spin much faster.

A slowly spinning cloud will spin faster and faster as it gets smaller and smaller. A spinning object that is formed by gravitational coalescence (such as a star) acquired its rate of spin from the initial spin of the cloud.

For a large enough mass, the gravitational attraction creates very high temperatures and pressures as the potential energy of gravitation is converted into kinetic energy when the mass coalesces. The atoms collide ever more violently as they rush together, the electrons are knocked off the atomic nuclei, and the mass becomes a plasma. That is, it is converted to a high-temperature gas of free nuclei and free electrons that cannot combine with each other, even though they are very close together, because they are moving with extremely high speeds. An example of this in our own solar system is Jupiter, in which the self-gravitational force creates just such a plasma of protons and electrons in its interior.

The pressure of a plasma can balance the inward force of gravity for any aggregate whose mass is less than 0.084 solar masses. If the mass is greater, then the temperature and pressure squeeze the nuclei so close together that hydrogen nuclei collide at very high velocities and fuse to yield helium. The mass starts to burn, lights up, and becomes a star. What happens next again depends on the initial mass of our aggregate. Stellar evolution is a balance between gravity, which tends to contract the star, and internal pressure, which tends to expand it.

The important point for our discussion is that stellar evolution can result in a white dwarf, a neutron star, or a black hole, depending on the initial mass of the coalescing cloud. These are all peculiar objects whose strange properties caused a lot of consternation in the scientific community.

The key to understanding compact objects and stellar evolution came from a study of white dwarf[69] stars. They were so mysterious and strange, and the ultimately correct theory describing them violated the conventional astronomical wisdom so much, that great minds came into dramatic conflict. The most distinguished astronomer of all, Arthur Eddington, refused to accept the theory, even after everyone else knew it was correct.

White dwarfs were observed long before they were theoretically understood. F. W. Bessel in Germany undertook a detailed study of Sirius (the Dog Star) in the decade of 1834 to 1844 and concluded that Sirius was part of a binary system. The hidden companion was called Sirius B, and it was first seen by A. G. Clark in 1862 while testing a new telescope.[70] He saw a small bit of light precisely where Bessel said it was supposed to be. The companion was very dim but no longer hidden to the increased telescopic power. The properties of the companion could then be obtained, and these were found to be completely baffling. Measurements of the star's spectrum showed its surface temperature to be 30000 kelvin, which meant that its internal temperature was millions of degrees.[71] Since the temperature had to come from the potential

[69] A white dwarf gets its name from the fact that it is very small and can be seen. Stars that are just as small but can barely be seen are called brown dwarf, and those that cannot be seen at all are called black dwarfs. Current theory states that the time to form a black dwarf is greater than the age of the Universe, so they are hypothetical objects.

[70] Since Sirius A is known as the Dog Star, Sirius B was dubbed the Pup.

[71] Stellar spectroscopic analysis consists of passing starlight through a spectroscope and determining the frequencies present. From the distribution of the intensity of the frequencies present, the surface temperature of the star can

energy of gravitational collapse, it must have condensed from a very massive object. Its orbit showed that its mass was just about the same as that of the Sun, but its temperature and luminosity showed that it was no bigger than a large planet, so its density is very great. This means that deep down in the interior of the star, where matter is crushed into small volumes by gravitational pressure, the density must be truly enormous, reaching values of a million grams per cubic centimeter, which is two hundred thousand times the average density of the Earth! A teaspoon of such stuff would have a mass equal to that of an iceberg two or three hundred feet in diameter and, if the Earth had the density of a white dwarf, its radius would only be about half a mile.

Eddington was the first to try and make sense out of this and suggested the atoms were completely ionized by the enormous temperatures and pressures inside a white dwarf so that all particles could be crowded closely together without the limiting effect of interatomic repulsions. But he could not explain why the star would not simply continue to collapse. The general view, shared by Eddington, was that the high temperatures in the white dwarf produced high internal pressure. Just as the pressure of an ordinary gas increases as it gets hotter, the extreme temperature in a white dwarf would produce huge pressures that would resist the crushing force of its gravitational field. But Eddington found that there was no way to construct a consistent theory on this basis and this posed an unsolved puzzle. The question was: how could the enormous gravitational field of such an incredibly dense star be resisted? Ordinary hot gas pressure could not do it, so what else was there? He could not solve the white dwarf problem because the solution depended on the quantum mechanical properties of electrons, and this knowledge was not yet available. At that time, the peculiar property of electrons called degeneracy was not yet known.

be calculated using the theory of black-body radiation. From certain specific frequencies, the composition at the star's surface can be computed, because we know the frequencies emitted by atoms at high temperatures.

The pressure of an ordinary gas, such as oxygen or nitrogen, is just the result of the kinetic energy that resides in the moving molecules and so depends in a simple way on the temperature and density. When the temperature of the gas approaches absolute zero, its internal pressure approaches zero. And when an external pressure is applied to an ordinary gas, its volume simply decreases to that of the molecules. This is not so for a gas of electrons.

A part of the resistance of an electron gas to compression is similar to that of an ordinary gas, but only a small part. By far the greatest resistance comes from quantum mechanical effects. A major consequence of quantum theory is that there are restrictions on the energies that gas particles can have. According to classical physics, the atoms or molecules in a gas can be distinguished from one another and can take on any energy whatever, with no limitation on how many of them can have the same energy. There is nothing to prevent two, three, or more molecules from all having the same energy. When the theory of such a classical gas is worked out, the relation between temperature, pressure, and density is easily obtained. And the theory implicitly assumes that molecules can be distinguished from one another. For the simplest cases, this yields the well-known ideal gas law.[72] Two quantum mechanical results destroy the accuracy of this classical result. The first of these is the fact that, at the quantum level, particles *cannot* be distinguished from one another. There is no way to paint numbers on atoms or electrons so they can be labeled. This simple fact has far-reaching consequences.[73] For electrons, there is a second, related fact. Only two electrons in an electron gas can have

[72] The ideal gas law holds for gases whose molecules do not interact. Even for gases in which there is an intermolecular energy of interaction, the ideal law is quite accurate at high temperatures.

[73] It was the implicit assumption of distinguishability that gave theoretical results for black-body radiation in stark disagreement with experiment. Planck's assumption of discrete energy levels for radiation amounted to bringing in indistinguishability, although he did not realize this at the time.

the same energy. This is the famous Pauli exclusion principle.[74] When the equation of state for an electron gas is worked out taking exclusion and indistinguishability into account, a result is obtained that is very different from that for classical gases. There is still a thermal contribution, arising from the ordinary kinetic energy of the electrons, but the pressure arising from the quantum effects is much larger. Since no more than two of the electrons can be in the same energy state, they strongly resist being squeezed together and exert a large pressure tending to force them apart. This is called the degeneracy pressure, because it comes from a gas of electrons that have all degenerated into their lowest possible energies.[75]

For high densities of electrons, the degeneracy pressure is enormous and quite sufficient to support the crushing gravitational field in a white dwarf.

Here was the first link between the physics of enormous objects (stars) and the quantum theory of fundamental particles (electrons) that has persisted to this day.[76] White dwarfs are stars whose existence is made possible by quantum mechanics.

In 1925 Walter Adams performed a definitive experiment, showing that Sirius B indeed had a small size and a large mass. He had first determined the star's surface temperature in 1915 by analyzing its light in a spectroscope. Eddington suggested that he look for a red shift in the spectrum of Sirius B because, if the star were indeed so massive and small, the light leaving its surface would be subject to a measurable general relativistic effect

[74] The exclusion principle actually states that no two electrons can be in the same state. An electron can have two orientations of its spin, each with the same energy, so two electrons can have the same energy in an electron gas.

[75] Complete degeneracy takes place only at a temperature of absolute zero where all electrons are in their lowest energy states. A white dwarf is very hot, but for high densities, electrons act as if they are at zero temperature even when their temperature is high.

[76] Modern cosmology shows that the nature of the universe on the largest scales, from stars and galaxies and beyond, cannot be understood without knowing the properties of the smallest subatomic particles, down to quarks and beyond.

in which the energy loss caused by gravitational attraction would show up as a red shift of all its frequencies. Adams looked for such a red shift and found it to be that predicted by Einstein's theory. More accurate measurements were carried out by Popper in 1895 on another white dwarf. These and later observations were convincing verifications of general relativity.

R. H. Fowler, in 1926, was the first to relate electron degeneracy to the properties of super-dense objects and he pointed out that the degeneracy pressure could support the gravitational force inside a white dwarf.[77] Subrahmanyan Chandrasekhar elaborated this idea in great detail in 1930 during an eighteen-day sea voyage from India to England on his way to Cambridge, where he had been accepted for graduate study. He was only nineteen years old.

Chandra had been exposed to the new quantum mechanics in 1928, when Arnold Sommerfeld visited Madras. He had read Sommerfeld's classic *Atomic Structure and Spectral Lines* and was anxious to meet the great man. Sommerfeld agreed to a meeting, during which he said that everything in his book was now outdated, and told Chandra about the new quantum mechanics. In the course of studying this fascinating new physics, he came across Fowler's article and Eddington's book on *The Internal Constitution of Stars*, so by the time he took his trip to England, he knew about the white dwarf mystery and about electron quantum degeneracy.

The results of those eighteen days at sea were remarkable. Fowler's work showed that the degeneracy pressure was important, but he had not obtained a complete description of the balance between degeneracy and the gravitational force, and he had not worked out the variation of gravity, degeneracy pressure, and density with depth in the star. It is noteworthy that in order to do this

[77] Fowler was one of the most distinguished physicists of his time. Much of his pioneering work on statistical mechanics is contained in two remarkable books, his *Statistical Mechanics*, and a second book, *Statistical Thermodynamics*, coauthored with Guggenheim. Several generations of scientists learned statistical mechanics from these books, and they can still be studied with great profit.

Chandrasekhar had to extend the theory of electron degeneracy to include relativity theory. At normal conditions, the velocities of the electrons in an electron gas are low enough that the classical relation between velocity and energy works fine. But the velocity increases as the electrons are crowded together, and at high densities their velocities are appreciable fractions of the velocity of light, so special relativity had to be included in the quantum theory of degenerate electrons. This combination of quantum mechanics and relativity became a focal point of Eddington's objection to Chandrasekhar's theory.

Relativistic velocities have a profound effect on the balance between gravitational force and the degeneracy pressure. In fact the effect is such that in massive white dwarfs electron degeneracy could not oppose the crushing force of its internal gravity. Chandrasekhar's theory stated that white dwarfs could exist only for masses less than 1.4 times the mass of the Sun. For larger masses, the star would continue to shrink. This was a troubling conclusion, and no one was quite sure what it meant.

The mass limit came from calculations combining the pressure–density relation for degenerate electrons with the stellar equations for the balance between gravitational compression and internal pressure. The calculations were easy only in the limits of small and of large white dwarfs, so the theory did not show that *all* white dwarfs had a mass below 1.4 solar masses. In 1934 Chandrasekhar performed the numerical calculations needed for stars of intermediate size and showed that the masses were *always* below the 1.4 limit. The work was long and tedious, because computers were not yet available, and the only mechanical help was that of a primitive calculator using gears driven by a hand crank. When he was done, Chandra[78] had shown unequivocally that the mass limit he had proposed held for all white dwarfs that could exist.

[78] He was thus called by his friends. Of course, I never knew Chadrasakhar, but I will refer to him as Chandra in the interest of brevity. The gracious and friendly nature of his character that history records leads me to believe that he would not mind.

He presented his results at a meeting of the Royal Astronomical Society on January 11, 1935. Eddington had arranged to make a presentation right afterwards in which he claimed that the mass limit result was wrong! If the limit were exceeded, the star would continue to shrink with nothing to stop it, and Eddington thought this ludicrous: a *reductio ad absurdum*, the existence of which proved that Chandrasekhar had to be wrong. Nature could not behave that way. A star simply could not shrink to nothing!

Eddington had said nothing to him about a disagreement, in spite of the fact that he had been watching Chandra's calculations almost on a weekly basis and was thoroughly familiar with his theory. The attack was a complete surprise!

Chandrasekhar was a young newcomer, and Eddington was the most distinguished astronomer in the world, so his view won out. In fact, since Eddington thought he was wrong, Chandra felt there was no future for him at Cambridge, so in 1936 he left for the University of Chicago. Remarkably, in spite of this shabby event, the two remained on friendly terms for the rest of their lives.

Ultimately, Eddington was proven wrong and Chandrasekhar right. Eddington's motivation was simple repugnance at a result that he felt nature would never tolerate. How could anything continue to shrink forever? What would be the final result? But he had to find some reason why Chandrasekhar was wrong. He thought he found it in the way Chandra had combined relativity and quantum mechanics to get the electron degeneracy pressure, and he proposed a different mode of combination that gave the answer he was so sure was correct. But the best quantum theorists rejected Eddington's method, and the mass limit for white dwarfs is now a standard part of the theory of stellar evolution.[79] The theory is supported by astronomical observation. No white dwarf

[79] The quantum theory of a non-relativistic electron gas yields a degeneracy pressure that is proportional to the 5/3 power of the density, while for a relativistic gas the degeneracy varies as the 4/3 power of the electron density. It is this relativistic result that leads to the Chandrasekhar limit. Eddington tried to prove (and failed) that the correct relativistic exponent was indeed 5/3.

with a mass greater than the Chandrasekhar limit has ever been observed.

What lay beyond this limit was controversial and unknown. Astronomers did not want to admit the possibility of a heavy star getting smaller and smaller indefinitely, but could think of no mechanism that could oppose gravity once the electron degeneracy was overcome. It took many years to fully resolve the issue, but as early as 1933 Fritz Zwicky had a bold leap of imagination that ultimately turned out to be true and pointed the way to go past the Chandrasekhar limit.

Soon after James Chadwick, in 1932, experimentally verified Rutherford's postulate that atomic nuclei contain neutrons, Zwicky started publishing his belief that stars existed consisting of nothing but neutrons, which he claimed were the end points in the evolution of massive stars. There was therefore no more paradox. Stars below the Chandrasekhar limit died as white dwarfs while heavier stars died as neutron stars. This was nature's way of avoiding the unthinkable shrinkage of a heavy star to nothing.

In a 1934 paper Walter Baade, an outstanding observational astronomer, and Zwicky proposed that when nuclear burning starts to die down in a massive star, the nuclear fire can no longer oppose its self-gravity, and the star implodes. The amount of energy released is enormous, and it drives an explosion producing the supernova. The gravitational force is so great that electrons penetrate the nuclei, combining with protons until all that is left is an assembly of neutrons. The explosion leaves behind a neutron star at its core.

The proposal turned out to be correct. In spite of this, and in spite of its being well supported by observation and calculation, it was largely ignored. A probable reason for this was that Zwicky was a prickly character who was certain of his superior intellect and treated other people's work with disdain. He was also prone to making outlandish speculations with little real evidence to back them up.

It was not until 1937, when Lev Landau wrote a paper claiming that neutron stars existed at the core of stars like the Sun, that the subject was taken up again. Landau was well respected and recognized as one of the world's great physicists, and the physics community eagerly read everything he wrote.[80]

After the Landau paper, a group led by Oppenheimer took up the study of neutron stars in earnest. The urgent question was: did or did not a maximum mass for neutron stars exist (analogous to the limit for white dwarfs)? If there were no upper limit, then all stars would die either as white dwarfs or neutron stars, and there would never be an object whose size could shrink to a singularity. On the other hand, if there were an upper limit, then black holes would form, violating the instincts of the vast majority of astronomers.

No one could have been better qualified to study this problem than J. Robert Oppenheimer. He entered Harvard in 1922 to study chemistry but came under the influence of Percy Bridgeman, one of the best experimental physicists of the time, whose study of materials under high pressures is still classic. Oppie, as he came to be known, then decided to pursue physics. Although reputed to be rather inept at experiments, he was a superb theorist and studied the emerging quantum theory with some of the best scientists in Europe, including Rutherford and Born. During that time, he published papers on quantum mechanics, and his genius was immediately recognized. In 1928, he joined the faculty at the University of California and founded the largest and best school of theoretical physics in the United States. More than anyone else, Oppie was responsible for bringing the new physics from Europe to America and creating a vibrant community of theoretical physics. He was interested in a wide variety of subjects besides science, especially

[80] Kip Thorne, in his fascinating book *Black Holes and Time Warps*, has poignantly described how Landau's neutron core paper was an attempt to avoid the Stalin purge of intellectuals. Although his work won international acclaim, it was not enough to deter the authorities, and in 1938, Landau was taken to prison. He was released a year later through the intervention of Kapitza. Landau's work spanned all theoretical physics, but his greatest accomplishment was his theory of superfluidity, for which he won the Nobel Prize in 1963.

Eastern philosophy and religion, and had a remarkable facility for languages. He had a fine aesthetic sense for literature, art, wine, and food, and he could indulge his tastes because of the inheritance left to him by a wealthy father. He was a charismatic leader who rapidly became widely known, and he was chosen by General Groves to lead the Manhattan Project. Oppie had no previous experience running large-scale organizations but turned out to be an outstanding administrator, and everyone believed that the successful development of the atomic bomb within a rather short time was largely due to him. After the war, his old left-leaning sympathies, combined with personal conflicts, led to the revocation of his security clearance in 1954.[81]

The critical missing information was the equation of state for stellar matter. This is the relation between the pressure, temperature, and density, and it must be known in order to calculate the pressure in the interior of the star that balances the force of gravitational collapse, The correct relation was known for material of lesser densities, but was poorly understood for the nuclear densities characteristic of neutron stars.

Oppenheimer's research strategy for the neutron stars was similar to Chandrasekhar's for white dwarfs, but with two major differences. First, while the classical Newtonian theory of gravitation was adequate for white dwarfs, it was not so for neutron stars. They were so dense, and the gravitational fields so intense, that the general theory of relativity was required. This made the mathematics much more complex. Second, and more serious, was the lack of knowledge of the nuclear forces. There are two sources of resistance to the internal gravitational pressure at very high densities. One is the degeneracy pressure of the neutrons, which is similar to that for electrons, but much stronger because

[81] This was the infamous McCarthy era in which a single Senator managed to do so much damage to American science. I remember a newspaper cartoon in which Oppenheimer was strapped to a chair with a steel helmet completely enclosing his head and a government bureaucrat shaking a finger at him. The caption was; "Now don't you dare think up any more government secrets".

the neutrons were much closer together; and the second is the nuclear force between the neutrons. Not much was known about this in 1938, so Volkoff, an Oppenheimer student who was doing numerical calculations, neglected the nuclear force and found that a neutron star must always have a mass lower than 0.6 that of the Sun. In the meantime, Tolman took a different tack by working out the theory with the dependence on the nuclear force as a parameter. He found that no matter what the nuclear force (within a believable range), the theory still gave a maximum mass beyond which a neutron star could not sustain itself. From attractive to repulsive nuclear interactions, the calculated mass was between one half and several solar masses. There was no way out. For large enough masses, the neutron star had to collapse to the singularity we now call a black hole. Modern estimates give the mass beyond which a neutron star collapses to a black hole to be between two and three solar masses. For large masses, implosion to a singularity is unavoidable. Nothing, not even neutrons, could stand up to the crushing gravitational force of sufficiently large masses of matter.

The first neutron star was seen in 1967 by Jocelyn Bell, using a radio telescope designed by her thesis advisor Anthony Hewish at Cambridge University. Soon after it was built, Bell detected signals indicating that a strong source of radio waves was being emitted every one and a third seconds. At first, the meaning of these results was unclear; the suggestion was even made that the precise periodicity of the signals implied they were from an alien intelligence. This was recognized as an outlandish idea and quickly discarded as many more such sources were found.

They were rotating so fast that they had to be very small.[82] In fact the angular velocity was so large that the object could only hold together if it had enormous strength. The obvious candidate was a neutron star. Later observations and calculations confirmed the existence of neutron stars. The neutron star was visible because it

[82] As a large, rotating object shrinks, its angular velocity goes up because it must conserve angular momentum. Rapidly rotating celestial objects are usually small objects that have condensed from larger objects.

had a large magnetic field that spewed out particles and radiation as it turned. Hot spots in the field produced regions of very high emission that spun around, giving off radio waves much as a turning lighthouse gives off light as rotating beams. A young neutron star has an intense rotating magnetic field that sets up a huge flow of electrons, positrons, and ions over its surface. This sets up a "pulsar wind" just like the solar wind from our Sun.

What strange creatures are spawned by gravity! The Sun and the stars are strange enough. When gravity condenses clouds of interstellar gas of sufficient size, the crunch converts so much gravitational energy to heat that the enormous temperatures become large enough to initiate nuclear fusion, converting hydrogen to helium and starting the life-to-death cycle of a star. A star is a marvelous thing in itself, with huge masses, surface temperatures of thousands of degrees, internal temperature of millions of degrees, and lifetimes of tens of billions of years. And the things that are produced on the way to stellar death are beyond imagination. White dwarfs are so compact that a mass equal to that of the Sun is squeezed into a volume equal to that of the Earth. Neutron stars are even denser, and they have a solar mass compressed into a volume whose diameter is merely that of a small city. The density is equal to that of nuclear matter, and in fact a neutron star is just like a giant neutral atomic nucleus.

A neutron star is strange enough, but the weirdest object of all is the black hole, which is so dense that it is squeezed into nothing!

19

Beyond existence

Gravity organizes the cosmos and controls the way it changes. And in the course of stellar evolution, the elements necessary for the formation of life are made. As the mass of an aggregate of hydrogen is made larger and larger, its internal structure finds it ever harder to resist the crunch of gravitation. First, the ordinary compressive resistance of atoms gives way, and the atoms ionize to form a plasma composed of electrons and ions. Large planets, such as Jupiter, are composed largely of just such plasmas. The conversion of gravitational to kinetic energy attending the addition of more mass brings high-velocity protons very close together; fusion reactions then convert hydrogen to helium, and a star is born. The high temperatures and densities required for fusion exist only near the center of the star because gravitational pressure in the star's outer layers is not high enough to produce such conditions. The star can go on for billions of years shining brightly with the radiation from the nuclear fire, but sooner or later the core runs out of hydrogen. The internal pressure in the core is then not strong enough to resist gravity, so the star contracts rapidly and becomes very hot because of the conversion of gravitational potential energy. The hydrogen in the outer layers then becomes highly compressed, fusion takes over again, the outer layers expand to a very large size, and the star becomes a red giant, leaving behind a massive core that is largely helium. The basic element needed for life is created at this stage, because the temperature and density in the core are now large enough to induce the nucleosynthesis that converts helium to carbon. The end of this process is a white dwarf.

If the mass is large enough and the temperatures are high enough, nuclei heavier than carbon and, in fact, all nuclei up to and including iron, can be formed. No heavier nuclei can be made because iron is the most stable nucleus at these temperatures and pressures. For a massive enough white dwarf, the end of the nuclear cycle produces a gigantic explosion. The star quickly collapses and the gravitational energy that is released is so great that the white dwarf is converted into a neutron star in a matter of *seconds*. The total energy of the explosion equals that generated by the Sun for its entire ten-billion-year lifetime, so the temperature increases enormously. The elements beyond iron in the periodic table are all made in these supernova explosions.

If the remaining core has a mass greater than the Chandrasekhar limit, but less than about two and a half solar masses, it will be a neutron star. For larger masses, it becomes a black hole.

White dwarfs were first observed and later explained by theory. For black holes, the opposite is true. They were first found by theoretical calculations and were looked for and found only after these calculations became convincing.

Michell's speculation was based on a simple result of Newtonian theory; namely, that the escape velocity increases with a celestial object's mass. From Newton's theory, it was easy to compute the mass for which the escape velocity equals the velocity of light, thereby giving what we might call a classical black hole. The mass was enormous, and, because Michel did not contemplate super densities, he thought its volume was also huge. With the corpuscular theory, light behaves just like a material particle so even light cannot escape from a classical black hole.

The modern theory of black holes really started with Karl Schwarzschild. He was only eleven years old when he became interested in astronomy, and his brilliance was apparent at an early age. By the time he was sixteen he had already learned enough celestial mechanics to publish two papers on the orbits of double stars. A year later he was interested in the geometry of space and speculated that physical space was non-Euclidean. In a paper of 1900, he discussed the possible lower limit on the curvature of space. This was long before Einstein started to work on relativity. He read Einstein's general relativity paper right after it was published in 1915 and immediately solved the field equations for the space outside a perfectly spherical, non-spinning object. A few weeks later, he found the solution for the space inside the sphere. Schwarzschild had found the first solution of the field equations of general relativity. He sent a paper on this to Einstein, who presented it to the Prussian Academy of Sciences on his behalf early in 1916. At about the same time, Schwarzschild published a paper on the quantum theory of the Stark effect (the splitting of spectral lines by an electric field). He wrote these papers while serving in the German army on the Russian front, where he contracted a disease of the immune system. He was dead from this illness by May of 1916. He was forty-two years old. Schwarzschild obtained the metric in the neighborhood of a sphere. That is, he had discovered just how the gravitational field of a massive object warps space and time. His result was simple and unequivocal.

Far from the object, gravity was very weak, so time behaved normally and space was Euclidean. But the gravitational field increased greatly closer to the sphere, and both time and space measurements changed drastically. Time was shortened by a factor depending on the mass of the sphere, and the contraction of time intervals increased dramatically as the distance from the sphere decreased. At the same time, distance intervals increased by the same factor. This was understandable and acceptable except for one disturbing fact. Close to the sphere, there is a distance at which all time intervals are zero! At a certain critical radius, time stops, so that any light at that radius must have a zero frequency, and its wavelength is shifted to infinity. The light is *red shifted out of existence*, at least from the viewpoint of an observer outside the sphere. The decrease in time intervals means that as an object approaches the critical radius, it moves ever more slowly (from the point of view of an observer outside the critical radius) until it stops at the radius and hangs there. But someone riding on the falling object sees his motion quite differently. In the reference frame of the moving object, time never stands still. Rather, time seems to speed up until it is flashing by. In his own frame, the rider falling to the Schwarzschild radius just keeps falling right to the center of attraction with a speed approaching the speed of light. The enormous gravitational field would induce such strong tidal forces that an object would stretch out in a radial direction, and compress in a tangential direction, so it would be infinitely thin and infinitely long. The center of the black hole is a singularity that crunches everything into nothingness! All the matter that goes into a black hole loses its identity. No matter what it was made of, hydrogen, neutrons, plasma, or anything else, its characteristics disappear. Everything is gone; except that is, for its mass. A black hole is a singularity that has eaten everything that fell into it, but it still has the mass of the matter that formed it.

The critical radius is called the event horizon because an outside observer can see nothing beyond this point. The gravitational field is so strong that the escape velocity for anything, matter or light is greater than the velocity of light. Once inside the event horizon, nothing can be seen and nothing can come out. The radius of the event horizon is proportional to the mass of the black hole and is given by a simple formula. Incredibly, this formula for the radius at which the escape velocity is the velocity of light is the same as that found by Michell. Both the Michell and the Schwarzschild calculations give the same answer for the size below which nothing, not even light, can escape the gravitating sphere. The answer is the same, but the physics is different. Michell's calculation, based on Newton's theory of gravity, states that a light corpuscle can leave the sphere and climb away from it, but it must ultimately fall back. This is just like a rocket ship fired from Earth whose velocity is below the escape velocity. In the Michell theory, an observer (using photocells perhaps) beyond the critical radius would therefore detect light going up and then falling back, just as a ball thrown into the air on Earth would rise and then fall to the ground. But the Schwarzschild theory is a result of general relativity, and an outside observer would see nothing leaving the black hole.

No one could accept this. Eddington and Einstein certainly couldn't. Eddington's objection was just like his objection to the Chandrasekhar limit: nature simply couldn't have such singularities. Einstein even wrote a paper in 1938 with a derivation that he interpreted as showing that the Schwarzschild event horizon could not exist. Many physicists did not think that such singularities would have any real consequences anyway. The event horizon for a mass equal to that of the Earth would have a radius less than a quarter of an inch, and for a mass equal to that of the Sun, the radius would be only about three kilometers. There was no conceivable way to pack so much mass into so small a volume, so they thought they were looking at a non-problem.

Readers who are looking at relativity for the first time and find its results unsettling should take comfort from this. Even the greatest geniuses will sometimes balk at concepts that violate their sense of physical reality.

The Oppenheimer–Snyder approach, in which the equation of state of nuclear matter was used to calculate the pressure that opposed gravitational force, was taken up again in the sixties. Many became convinced that, for a large enough mass, implosion is unavoidable and therefore black holes must exist. A lot of research followed in which the properties of black holes were worked out, and eventually observational astronomy confirmed their existence.[83] There is now strong evidence that super-massive black holes exist at the center of several galaxies, including our own. Perhaps most galaxies have a black hole core.

They certainly are strange; but the most important scientific result is that black holes bring the two great modern theories of physics into direct contact. Relativity is the science of large gravitational fields, and quantum mechanics is the science of small particles, so they meet at the center of a black hole, where gravity

[83] There are a number of excellent books on black holes that describe them in detail. Most noteworthy are those by Kip Thorne and Stephen Hawking. My purpose here is merely to point out the inevitable consequences of gravity and the strange, non-intuitive objects it creates in the universe.

approaches infinity and sizes approach zero. While the study of black holes has not yet resulted in a marriage, some coupling of the two theories has been successful.[84] If nothing else, it has pointed theory in certain directions and has certainly enhanced our understanding of black holes.

[84] For example, Stephen Hawking has applied quantum theory to black holes to show that matter *can* leave a black hole by quantum mechanical tunneling. The rate is slow, but it exists.

20

Absolute space?

In 1963, two young astronomers at the Bell Telephone Laboratories heard some static that told us how the universe began. Arno Penzias and Robert Wilson were trying to use a Bell radio telescope to study the Milky Way, but could not get a decent signal because of the presence of some persistent radio noise. They thought that the most likely origin of the problem originated with some annoying pigeons whose droppings fouled the radio antennae, but they found the static continued even after the most careful cleaning. Careful observation showed that the noise signal did not originate in the Milky Way, or from any specific point in the sky, but came uniformly from every direction. They knew that Robert Dicke at Princeton had predicted that such radiation should be a relic if the universe indeed had started with the Big Bang, which was the favored cosmological theory. They called Dicke, who visited them, looked at their equipment, and discussed their observations. They agreed that the best explanation was that the static was indeed the leftovers of the Big Bang.[85]

[85] Dicke was not the first to point out that the Big Bang should leave a residue of low-level radiation in its wake. Georg Gamow and others had made this prediction much earlier. In fact, calculations of the "temperature of space" had been made even in the nineteenth century, on the basis of theories that had nothing to do with the Big Bang, so a uniform distribution of radiation in the universe was not a new idea. But Dicke's work was the most detailed, and in fact he, with David Wilkinson, was in the process of building a radio telescope precisely to look for such radiation. When he learned of the Penzias–Wilson results he said, "We've been scooped".

The origin of the universe has fascinated people ever since the beginning of human consciousness and it is still a compelling issue today. Specific questions are: Is the universe infinite or finite? Did it exist forever or did it have a beginning? Will it last forever or will it end? Let's pursue these as scientific rather than theological questions; that is, we look to observation and experiment for answers.

The idea that the universe is infinite produces some contradictions that were recognized very early in the era of modern science. Newton himself saw that a finite universe could not have the distribution of stars we observe, because gravity would have caused all matter to coalesce into one great mass long ago. A German astronomer, Heinrich Wilhelm Olbers, popularized an argument for a finite universe in 1823, and, although discussed as early as 1619 by Kepler, it became known as Olbers's paradox. The paradox is simply described. If the universe is infinite, then any line-of-sight direction to the sky will intersect an infinite number of stars. Therefore, the sky should be very bright at night instead of being a dark field with points of light spread on it. The puzzle was solved when Edwin Hubble examined the spectra of a number of galaxies at various distances from us. He found that all the spectra were shifted in frequency toward the red end of the spectrum and that the further away the galaxy, the greater the shift. This is interpreted in terms of the well-known Doppler effect in which the velocity causes the frequency of a receding light source to decrease. The velocity of the galaxies relative to us could therefore be computed, and this revealed one of the most important and astonishing results of twentieth-century astronomy. *The galaxies are all receding from each other at rates that increase with increasing distance between them.*

The Doppler effect shifts the light we get from the galaxies to the red, and as the distance of a galaxy from us increases towards infinity, the light becomes red shifted out of sight. Furthermore, the expansion implies that at some time in the past all galaxies were extremely close together and that some event caused them to mutually fly apart. This is the essence of the Big Bang theory.

The number of *visible* galaxies in the universe is therefore not infinite and there is no Olbers's paradox.

Detailed information on the nature of the background "noise" found by Penzias and Wilson was obtained from a satellite launched by NASA in 1989 called *COBE* (*Cosmic Background Explorer*). It is called the cosmic background radiation because it exists everywhere and has no single source.

The entire universe is bathed in a sea of cool radiation, practically all of it consisting of microwaves. The frequency distribution of this radiation is the same as that of a black body whose temperature is 2.73 kelvin above absolute zero. All evidence shows that the cosmic background radiation is left over from the origin of the universe. The Big Bang theory states that the enormous temperature in the primordial event about fifteen billion years ago would have cooled down to just 2.73 degrees absolute today as the universe expanded from the initial singularity to its present size. The existence of the background radiation, and its black-body temperature, is one of the major reasons for accepting the Big Bang theory.

The *COBE* measurements are extremely precise and show that the cosmic background radiation is almost the same for every direction in space. That is, the equivalent black-body temperature is the same everywhere in space within a relative temperature difference of less than 10^{-5} degrees. No existing single source can be responsible for such incredible uniformity. Nevertheless, the small measured fluctuations are important because they reflect the early irregularities in the universe that resulted in the formation of galaxies. If the early universe were *perfectly* uniform, the present distribution of matter would be perfectly uniform, and galaxies would not exist because there would be no regions of differential density that could coalesce and separate from the rest of the universe. Even if some fluctuations in density did arise because of statistical variations of particle velocities, the distribution of galaxies would be homogenous throughout space. On the contrary, astronomical observation reveals a large-scale structure in the distribution of galaxies that is far from uniform. In 2001 the

NASA satellite probe *WMAP* (*Wilkinson Microwave Anisotropy Probe*) was launched to measure the background radiation over very large regions of the sky. This showed that fluctuations in the microwave density, and therefore fluctuations in the deviation of the early universe from perfect uniformity, existed that could account for the large-scale structure of the universe.

One fascinating result of the measurements is that there is a slight color shift in the radiation that varies with direction. It is blue shifted in one direction and red shifted in the opposite direction. This is a Doppler shift from which the velocity of the Earth (actually the velocity of our local galaxy) can be calculated relative to the cosmic background radiation; that is, relative to a coordinate system anchored to the general expansion of the universe. The velocity is about 600 meters per second and is moving us towards the constellation Leo.

This presents a most interesting possibility. Since the coordinate system tied to the expansion of the universe is everywhere the same, and since the microwave background can be used to measure velocity with respect to this system, why cannot it be used to define a privileged frame of reference and therefore give us an absolute space? Does a privileged system exist and can it be used to define the absolute space of Newton?

Actually, the background radiation *can* be used to define a special frame of reference which is the same for everyone and indeed it can result in universal velocity measurements. But this is not the same as saying it defines a Newtonian absolute space, and it certainly does not negate Einstein's relativity. The relativity principle states that the fundamental laws of physics are everywhere the same, and these are given in terms of derivatives and rates that are independent of location. Just as the laws of hydrodynamics do not change from place to place because of the surrounding fluid, so the laws of physics do not change because of the existence of a cosmic radiation. Such radiation can be used to define a coordinate system, but it is just one of an infinite number of coordinate systems that can be used to describe physical results. The same can be said for the finite lifetime and finite size of the observable universe. Just because there was a Big Bang with a special point in time (a beginning) does not mean that the laws of physics are not everywhere the same. Indeed, if there is a beginning of the universe, then by definition, there certainly was a beginning to time and space, and the laws of physics did not apply before that moment. But the principle of relativity for all *accessible* places and times still applies.

Let's pause on that very important clause in the last paragraph. If there was a beginning of the universe, then *by definition* there was no time and space before that moment. The reason for this is that physical time and physical space are what we measure with clocks and rulers. If these do not exist, no measurements can be made, so physical space and physical time cannot exist.

Similarly, an "absolute" space can be defined by the quantum vacuum. According to quantum field theory, the vacuum is not really empty. It is full of virtual particles that continually break into reality, live for exceedingly short times, and then recombine back into the vacuum to nothing. For example, an electron–positron pair might suddenly appear and very quickly recombine. The effects of such virtual particles have been observed, and their existence is well established. An "absolute" space could be defined as one in which all virtual particles are at rest.[86] But again, this is not absolute in the Newtonian or relativistic sense. It is still true that the laws of physics are the same in all coordinate systems.[87]

The microwave background radiation, the possible beginning or end of the universe, and the existence of the quantum vacuum do nothing to negate the validity of general relativity. The reason for this is that relativity is merely a theory of what it means to make measurements of distance and time.

[86] More accurately, the absolute space of the quantum vacuum would be that in which the average displacement of all virtual particles would be zero.

[87] Quantum theory is not treated in this book, and it therefore contains no explanations of quantum phenomena. Suffice it to say that the results of quantum mechanics are much stranger than those of relativity and much more difficult to picture. Feynmann once remarked that any physicist can understand relativity, but that no one understands quantum theory.

Now let's reconsider the concept of an ether. It was originally invented because scientists believed something had to carry the forces acting between distant portions of matter, and with the success of Maxwell's theories, *something* had to carry light waves. But in its original form the ether had to be given up because it had self-contradictory properties and there were no detectable experimental consequences of its existence. But modern theory gives a new picture of what we use to call "empty space". Fields, curved space, and virtual particles assign properties to the spaces between objects that satisfy the original reasons for postulating an ether. There is a medium that carries the forces. We just don't give it the old name, but we can call it the "modern ether" or the "field/quantum ether" because it fulfills a similar function. There is, of course, a major difference from the old concept. The old ether was undetectable, while the "modern ether" is made experimentally visible by, for example, electric or magnetic probes. It has a physical reality absent in the old ether.

We have to accept another strange aspect of nature: "empty space" is not empty.

21

Infinity

The efforts to understand gravitation led to the theory of general relativity, and this ultimately poses profound questions about the universe. The deepest and most fascinating of these are about the extent of the universe in space and time. Did it exist forever and will it continue to exist forever? Does it extend forever in every direction? Or did the universe have a beginning, and does it have an end? And if we keep traveling out into space, will we come to an end?

These are questions that, if we try to think about them seriously, send our minds reeling. Yet there actually are some observations, and plenty of theory, that have something to say about them. The most important data are the Hubble expansion and the cosmic background radiation. Hubble's initial measurements, which showed that all galaxies are receding from each other, have been confirmed and extended, and expansion is now taken as an established feature of the universe. This, along with the cosmic background microwave radiation, leads to a beginning for the universe in the following sense. If we think of the Hubble expansion in reverse, the universe clearly goes to states of higher and higher density as we go back into the past because the galaxies were closer together long ago than they are now. Go far enough back and the size of the universe is very small, in fact approaching zero, and the density becomes extremely high, approaching infinity. So there was a beginning. At some instant, about 14 billion years ago (time zero) the super-small, super-dense universe started to expand and evolved into the universe we see today. With careful attention to the details of the expansion and to the nature of fundamental particles possible at various temperatures, this notion is the current Big Bang theory of the universe. It is remarkably successful. Not only does it account for astronomical observations, but, it also correctly predicts the formation of the elements and their relative abundances.[88] But it still leaves some important questions unanswered, such as the nature of physical laws at the origin of the universe and even of the existence of time, space, and physical laws before the Big Bang. And do such questions have any meaning? These issues are at the forefront of research on theories that try to unite quantum mechanics and general relativity. Without an understanding of quantum gravity, there is no hope for answers.

But we are convinced that there was a beginning. Will there be an end? Gravitation lets us at least examine the possibilities.

All galaxies are rushing away from each other because of the initial expansion, and we might expect that they will continue to do so forever. Then there would be no end, just a continual rarefaction, so there would be more and more empty space with stars and galaxies ever further apart. But the force of gravity opposes the continual expansion. Remember that gravity acts always and everywhere, so all the matter in the universe is mutually attracting, and this is a force that tends to coalesce everything into one great mass. The end result depends on how much matter exists. The greater the density of matter, the more powerful the mutual pull of gravity, so, if there is enough matter, the expansion of the universe will eventually slow down and reverse. Everything will collapse into a very small volume and very high density, beyond even that of a neutron star. There will be a Big Crunch, which is just the opposite of the Big Bang. An attractive scenario is that of a cyclic universe in which the Big Crunch leads to a black hole which somehow becomes the origin of a new Big Bang and so on.

[88] The beginning was not a gigantic explosion as implied by the name of the theory. Rather, it was an expansion that was initially indeed very rapid, but nonetheless smoothly continuous right to the present day. The theory was labeled “Big Bang” as a derogatory term by those who opposed it.

Below a critical density, the force of gravity is not enough to oppose the expansion, and the galaxies will recede from each other forever at an increasing rate. This would result in "heat death", an idea that was popular in the nineteenth century. Thermodynamics tells us that all energy will eventually be evenly distributed. The stars will exhaust their nuclear fuel, gravitational energy gradients will be ineffective, and the end will be a cold, dark universe containing only low-level radiation and the ashes of once glowing, fiery matter.

Clearly, there is a critical density of matter in the universe, above which everything must collapse to the Big Crunch and below which the universe must continually expand as it is doing now.

At this critical density, there will be continuous expansion, and until recently it was believed that the expansion rate would get slower and slower as time went by. Observations of the Hubble expansion in the past decade, however, have shown that the expansion rate is speeding up.

Astronomers have tried to estimate the overall density of matter and, in the past, have always found the density to be small, so the best guess was that the universe was open, in the sense that it would never fall back on itself. But as time passed, the estimates gave larger and larger values for the density, and now there is a general sense that the density is very close to the critical value, so the universe will continue to expand, but at an ever decreasing rate.

Two factors invite attention. The first is that observational inadequacies will always show a smaller amount of matter than actually exists. Telescopic techniques rely on electromagnetic radiation, and even though their range has been extended from optical to radio to infrared frequencies, there is no guarantee that we can see everything that is out there. Matter that does not radiate cannot be seen. The second fact is that a universe with exactly the right amount of matter to have the critical density is a remarkable coincidence and cries out for some explanation. The critical density is a "magic number", and magic numbers just don't happen by accident. There is general agreement among cosmologists that observational evidence strongly supports the notion that the universe is indeed flat. The flatness did not fit in well with earlier theories, and the Big Bang was therefore modified by postulating an enormous, and enormously rapid, inflation of the universe almost immediately after its beginning. The inflation theory results in a modified Big Bang model that is extremely successful in accounting for what we see.

There is a fascinating point that merits more attention. Modern measurements of the Hubble expansion show that the rate of the recession of galaxies is faster for galaxies that are further away. It's as if some force is pushing the galaxies apart at a rate greater than that arising from the initial Big Bang. In 1917 Einstein postulated just such a force when he applied his general relativity to a study of the structure of the universe. He ran into a problem because the equations predicted that no stable universe could exist. The only way to get around this was to add a term to his equations that acted to oppose gravity. This term is called the cosmological constant. But Einstein's real problem was that he assumed that the universe was static. General relativity gives a perfectly good description for an expanding universe, and when Hubble demonstrated the expansion, Einstein called the cosmological constant his greatest blunder.

But modern observations of expansion rates indicate that the cosmological constant might exist after all, and that it may be related to the energy of empty space as described by the quantum mechanics.

Still, this does not alter the conclusion that the universe had a beginning.

Only a cyclic universe, in which every Big Bang is followed by a "Big Crunch", avoids the question of the beginning or end of time. It had no beginning and will have no end: it always was and always will be. But in an ever expanding universe, while time will never end, it can at least be defined to have a beginning by saying that it started with the Big Bang. Observationally, there is no direct way of telling if this is actually the case. Cosmology is different from the other sciences in a very important respect. It is a historical analysis and tries to ascertain what happened in the past by observations of what exists now. Always, this involves a chain of inference that cannot be tested directly. A cosmological theory is accepted as correct if it yields deductions about *the present* that are in agreement with observation. Big Bang cosmology does indeed agree with what we now see, so we accept its content. Note that this is a considerable extension of the idea that science is based only on observable and reproducible facts. But it works, so we use it. In this sense, then, we say that time had a beginning and that it started with the Big Bang. The deduction that the universe had a beginning raises a host of extra-scientific questions that have great philosophical and religious significance. But this is neither an argument for or against the Big Bang theory. If the scientific evidence had led to the conclusion that the universe always existed just as it is now, the same kinds of questions would arise.[89]

What about the size of the universe? One way to define its size is to estimate the maximum distance we can see out into space. A theoretical limit is given by the Hubble expansion. Because of the Doppler effect, the frequency of light vanishes as the velocity of what we observe approaches the speed of light. From the known value of the Hubble expansion, we know that this corresponds to a distance of about 13.7 billion light years. Anything beyond this, if there is anything beyond this, is not visible to us. So we might take the universe to have a diameter of 27.4 billion light years. However, this does not take into account the fact that the Universe was expanding while the light was traveling towards us. Factoring in this expansion gives 156 light years for the diameter of the Universe. This is big, but it isn't infinite. It's also not very satisfying because we didn't answer the question of the size of the universe, we only made a calculation based on how much of it we can see.

[89] In fact, just such a theory has been proposed and seriously studied by astronomers and cosmologists. This "steady state" theory was rejected on observational grounds.

This is all rather freewheeling and has the aura of the "gee whiz" science that is entertaining but seldom informative. It's time to get back to basics and think about what we mean by infinity.

The word "infinity" has many connotations that relate it to mystery, impossibility, and the supernatural. But its use in science must be attached to physical reality and be capable of an operational definition. The easiest place to start is with mathematics by asking a simple question: What is the meaning of infinity in simple arithmetic? Is there an infinite number? Everyone knows that a count of integers can go on forever, and in this sense there are an infinite number of integers, but it is useful to put this fact in more formal language as follows: for any integer, no matter how large, a greater integer can be found. Choose an integer, say one million; then by simply adding the integer "one" we get a larger number. If we had chosen the integer a million million or even a billion billion, a larger number could be constructed just by adding some integers to our choice. This is the full meaning of the looser statement that the number of integers is infinite. This simple fact is the basis of the meaning of infinity in mathematics.

Similar considerations hold for physical quantities such as length and time. So here is the definition of an infinite universe. If for any measurement of length, another measurement can be made that gives a larger length, then the universe is infinite in space. Similarly, if for any measurement of time, a larger reading of the clock can be found, then we say the universe will exist forever. This is the total meaning of the ideas of infinite space or infinite time and is based on the actual physical operations of measurement.

Notice that there is a real difference between the concept of infinite space and infinite time even after we take into account their relationship as required by relativity. Let's accept the optical methods of determining distance that have been developed by astronomers. They are the best we have for defining astronomical distances. But because of the Doppler shift from the Hubble expansion, there is a limit to the possible distance that can be measured. Observationally, we are limited to measuring distances below those for which the Doppler shift goes to zero frequency.

The operational answer to the size of the universe is clear. *The universe is finite and has a diameter of about 25 billion light years.* This statement is bound to provoke doubts and questions. What exists beyond the visible edge of the universe? Just because we can't see it doesn't mean its not there; that would mean that reality is defined by what we see, not by an objective nature. What would we see if we actually traveled 12.7 light years to the edge of the visible universe? These questions cannot be answered.

Time is different because to find out if there is a time such that another interval of time could not be added to it, we would just have to wait. But similar considerations apply. Physical time is infinite only if for every measurement of time, another measurement can be made that gives a greater value. Our experience tells us that in this sense, time is infinite; all measurements of time we ever made have been greater than previous ones. Of course, we do not know what future measurements will tell us, so from the point of view of rigorous physics, based on what we can measure, we do not know if time is finite or infinite, only that if we extrapolate the evidence we have so far, time is probably infinite.

22

How weird can it get?

The scientific description of the physical world stemming from the theory of relativity is not merely strange. It is utterly bizarre! Light that always travels at the same speed no matter how fast we are moving; rigid rods that shrink and clocks that slow down just because they are in motion; mass and energy being the same thing; light being bent by gravity; curved space and time: all this sounds a lot more like magic and fantasy than science. The phrases themselves evoke a mystic feeling of the kind we associate with mystery religions, or the occult. Yet the facts, ranging from astronomical observations to nuclear bombs, have verified these conclusions, so it is difficult to deny them. Can the world really be that weird? Is it possible that physical reality can be so much at variance with common sense?

Yes it can! Remember that each step taken on the road to these strange conclusions was in accord with common-sense logic and actual observation. Sometimes a leap of imagination was necessary, such as when Galileo constructed a whole new picture of the universe from looking through his telescope or when Newton decided that celestial and terrestrial gravity were the same, or when Einstein assumed that physical laws were the same in every coordinate system, not only inertial ones. And sometimes it takes genius to go wherever experiment leads, as when Einstein stubbornly took the constant velocity of light to its ultimate conclusions. But these "educated guesses" were extensively checked against observation and experiment before they were accepted as true. At every stage, nothing contrary to ordinary logic was brought in. This common-sense, logical approach gave us the strangeness.

Weirdness is a mental phenomenon. It comes from the sense of what is normal and what is not that we construct in our minds from the environment we are born into and the experiences we have over our lifetimes. That environment and those experiences determine what we think of as normal, and anything else is deemed strange and alien. It is only familiarity that causes us to unconsciously accept the world we see as natural and ordinary. Yet, if we look carefully at what we see, and if we try to strip away the effect of our daily encounters with the physical world, we find an array of profound mysteries as perplexing as anything in the most esoteric reaches of science. The ordinary gravitational attraction we feel every day is no less awesome and no less beyond understanding than the results of general relativity. The usual manifestations of light, from rainbows to shadows, are just as mysterious as the fact that its velocity is constant. And the very existence of matter is incomparably stranger than the fact that rigid rods get shorter when they are moving.

The results of relativity are thought to be peculiar because they are not part of our daily experience, which is full of phenomena that would seem just as odd except that we see them every day. Relativity becomes visible only for the very large and the very fast. It shows itself only under extreme conditions. For terrestrial velocities, and for masses that are smaller than those of stars, the effects of relativity are trivially small and generally unobservable except with the most sophisticated and delicate of scientific instruments. Of course relativity seems weird. We didn't grow up with it.

At another level, weirdness depends on what we traditionally accept as reasonable scientific explanations. Galileo's laws of falling bodies were based on accurate measurements and sound logic, but many natural philosophers found them too strange to believe. For centuries, they had lived with a description of falling bodies in which bodies fell because it was the nature of heavy bodies to seek the center of the Earth, and the heavier the body, the more eagerly it tried to get down. The idea that every object falls at the same rate, irrespective of its mass, ran counter to generations of mental conditioning, so Galileo's ideas seemed strange and contrary to reason. Similarly, the Copernican heliocentric solar system, with its spinning and revolving Earth, was thought to be nonsense by those reared in the Aristotelian school of thought. A rapidly spinning Earth, careening madly around the Sun, was contrary to everyday observation; it was weird and silly.

By the twentieth century, scientists had learned that common sense based on traditional knowledge was not always a reliable guide, and the only way to decide on scientific truth was through reproducible experiment and observation. The two great revolutionary developments, relativity and quantum mechanics, were accepted with surprisingly little opposition, even though both theories radically violated some of our most cherished notions about space, time, causality, and the nature of matter. The exceptional case in point was Einstein's refusal to accept quantum theory as the final word. Quantum theory is statistical with a non-causal random element at its core. Niels Bohr was a pioneer in the creation of quantum theory and insisted that it was required by experiment, and he argued with Einstein over this for decades. Bohr held that if experiments said that nature at bottom was unpredictable, so be it. Einstein did not. Even his great genius could not break loose from his profound philosophical conviction that nature is not random and must be explicable in causal terms. But note that he did not reject the results of quantum mechanics; he only insisted that some deeper, causally based theory, must be behind it. This great debate has never been truly resolved, although there is a consensus that Bohr was right and that any single great theory that unifies all we know about nature will have a quantum foundation.

The essential point is this: however weird the results of our probes into the nature of the world, however strange and alien we find them, if we do the science correctly, it will tell us what nature is and how it works. The weirdness must be accepted as reality.

23

Scientific truth

Scientists seem to be satisfied only when they can identify chains of causality. But this is really just a psychologically satisfying shorthand way of describing science, because ultimate reasons are outside its scope. Yet there must be some connection between scientific results and reality.

For scientists, facts are the bedrock of reality. Objective, reproducible data are taken to *define* reality. From simple observations, such as "the Sun exists", to detailed, sophisticated measurements, such as those for the velocity of light, facts are the ultimate, incontrovertible real things in nature. That is why so much thought and expertise is put into finding the facts. They must be gotten right or there is no hope for understanding.

But no scientist believes that experimental results, no matter how accurate, constitute the ultimate goal of science or, by themselves, provide a satisfying picture of nature. It is the theories that can be constructed from the data that are important. Experimental knowledge is essential, but it is a stepping-stone. Faraday's experiments certainly were a great advance, but it was his concept of lines of force and Maxwell's field equations that provided a picture of what electricity and magnetism are and how electrodynamics works. The fields are taken to be a reality, so we must ask if that is really the case. Like true mystics, we are after the ultimate knowledge, so we must ask to what extent the entities we put into theories are really a part of nature.

I think the best we can do is the following: if a theory is believed to be correct, then we believe that for every element of the theory, there is a corresponding element in nature such that the relationships among them are the same as among the elements in the theory. Let's look at the theory of gravitational attraction. The elements in that theory are those of Newtonian mechanics: mass, inertia, force, distance, velocity, and acceleration. Because of enormous successes, we *assume* that these elements really exist in nature. More accurately, we postulate that there are real things in nature that are related to each other just like the Newtonian elements are related in Newton's theory of gravity.

Electrodynamics is a more striking example. The basic electrodynamic entities are electric charges and electric and magnetic fields. These were invented to account for experiment and defined entities that have a certain relationship to each other. Because of the success of the theory, we believe that things actually exist in nature that are related to each other in the same way as dictated by Maxwell's equations. This is the scientific approach to the reality of nature. It is a set of entities and relations that have the same function and structure as the elements of our theories. The search for ever better theories is a search for an ever more accurate description of the fundamental structure of the physical world.

It is possible to go deeper and ask for the process by which gravity or an electric force acts. The conflict between action-at-a-distance and action through an ether is an old one and implies very different pictures of how gravity works. Which one is correct? As always, the answer should start with what is actually observed, not only in the bare experiments, but also with respect to the theory. The action-at-a-distance picture was worked out in Newton's *Principia*, and the resulting mathematical scheme has been so successful that similar mathematical methods were developed for electrical and magnetic interactions. The ether picture allows one to assume that the only real interactions are those in which something happening at a point in space affects only the immediate neighborhood of that point. Maxwell showed that this concept leads to a mathematics that looks very different. I have already commented on the remarkable fact that theories of action-at-a-distance, and action through a local field (or an ether), are mathematically completely equivalent. The formalism for one scheme can be easily transformed to the formalism for the other. One gives a set of integral equations, while the other gives a set of differential equations. Calculus shows us how to convert one to the other. Either of two contradictory pictures of the deepest physical reality is totally consistent with experimental facts and quantitative description! Which is correct? In practice, theoretical physics has adopted a methodology in which field equations play the major role, and while the conversions from differential to integral descriptions are often used, the thinking of physicists is guided largely by the idea of fields. There is a power of prediction, as well as conception, in the local field concept that does not exist in the global, integral formulation. Differential equations can be solved once boundary conditions are given, and this allows calculations to be made for a wide variety of different cases. There is nothing odd about boundary conditions. They merely state the conditions for which a phenomenon exists. For example, to investigate the motion of an object using Newtonian mechanics, some specifications must be given. What forces are acting on the object? What is its initial position and velocity? If these are given, then the simple differential equation of Newton's second law can be solved to get the motion of the object for all time. This is how the orbits of planets are obtained. Knowing that the force of attraction between a planet and the Sun is given by the universal law of gravitation, and starting from some initial point in the orbit at a given time (both of which can be measured), the path of a planet around the Sun can, in principle, be calculated for all time. Similarly, the trajectories of projectiles or rockets are readily computed from their initial positions and velocities and the forces acting on them, using the same differential equation. The second law in differential form covers an enormous number of special cases.

The same is true for electricity and magnetism. The differential equations that constitute Maxwell's theory can be solved to describe the multitude of cases that arise in circuits, free charges, and the propagation of light. It is all so much easier than an integral approach.

The idea of an ether, as a substance that permeates all space and carries all physical interactions, has been rejected, but the basic concept has won out in the form of fields. There is nothing contradictory in the concept that local field effects lead to forces that can act over large distances. The opposite idea would be that bodies exert forces on each other with no intermediate agency, and the field equations simply describe this for small interaction distances. This is less satisfying, but no less mathematically correct.

Einstein's work changes this only in that it makes field theories more compelling. It is interesting that special relativity dispensed with the classical idea of the ether as a substance, and general relativity dispensed with the idea of gravitation as a force. After Einstein, there was no ether to carry any forces from one place to another, and no forces that had to act either locally or over long distances. Gravitating bodies did not move because they were subject to forces; they moved because they were following the shortest possible paths in a curved space-time. Motion, including accelerated motion, is just the inertial property of matter in a space that is non-Euclidean! There is no action-at-a-distance. General relativity is the ultimate field theory. All that counts is the point-to-point geometric properties of space-time. There is nothing else.

Apparently, the truth behind the observable complexities of gravitational phenomena, from Newton's apple to the cohesion of stellar galaxies, is curved space-time. It is so simple that it must be true. But it is important to repeat the nature of this truth. The experimental facts are manifestations of the primary physical reality, but the statements in the theory have a different relation to nature. Because the theory works so well, we assume that for all elements of the theory, there actually exist things in nature that correspond to them. Thus, in the theory, there are two essential, and related, theoretical constructs: the covariance of physical laws and the curvature of space-time. A rigorous epistemological statement would only say we *assume* that nature has a structure such that covariance and curvature defined in our theory can be mapped onto that structure. The practical physical interpretation goes further in that it takes this rigorous statement as just a discussion of the meaning of words. Certainly covariance and space-time curvature are physically real; if nature has elements that can be mapped onto the theory, why not give those elements the names they have in the theory? All physical theories are related to nature in this way. The degree to which we accept them as real depends on the degree of success they have in describing physical experiments and observations. It is in this sense that science is the continuing search for truth.

In all this, the equivalence of alternative verbal descriptions must be recognized. The failure of experimentalists to detect the ether was originally thought to be the result of the Fitzgerald–Lorentz contraction. That is, measuring rods were taken to actually shrink in the direction of motion. Length was taken to be invariant; matter itself was squeezed together by the motion. This, rather than the process of measurement, can be taken as fundamental and used to construct a consistent theory in agreement with experiment. Similarly, general relativity can be interpreted either as the effect of masses on space-time curvature, or as the effect of gravity deforming the paths of objects in a flat space. All physical theories can be verbally interpreted in more than one way. Curved space-time can be thought of as the modern form of the ether, as a medium that transmits forces and supports light waves.

So why do we adopt the Einstein formulation of special relativity rather than assume that matter is actually shortened by its motion? The answer is that Einstein's approach has several appealing characteristics that other theories do not. First, its assumptions are very few; second, it unifies mechanics and electrodynamics by having them both obey a simple transformation law; third, it dispenses with an unobservable entity (the ether); and finally, it provides a technique by which a huge number of physical situations can be understood, all from the same starting point. General relativity is interpreted as a curvature of space because it is a unifying concept, satisfies the simple, universal principle that the laws of nature should not depend on how we look at them, and gives the mathematical machinery to describe enormously varied and complex phenomena in the universe.

An essential point is that the consequences of the theory are the same no matter which verbal interpretation is adopted. Thinking of electrodynamics in terms of local fields or in terms of action-at-a-distance doesn't change the observable consequences of Maxwell's theory. Some scientists have therefore adopted the idea that there is *no* legitimate scientific description of physical reality. All that matters are the equations.

This is a logically consistent way out, but it is a lazy way out. It amounts to giving up what we wanted from science in the first place. A better and more satisfying approach is to recognize the equivalence of various interpretations. The wave–particle duality and the complementarity principle in quantum mechanics are both well established and have taught us that there can be more than one correct way of describing nature. Gravitation as a force, and as a curvature in space-time, are both equally correct ways of looking at the underlying reality; local fields and action-at-a-distance (as long as it is propagated at the speed of light) are equally valid ways of describing electrodynamics. The complementarity of physical interpretations may violate our Aristotelian prejudice for "either–or" solutions. Yet we know, from our deepest and most completely verified physical theories, that different aspects of nature can actually coexist, even though, according to common sense, they are mutually exclusive. The fact that scientific theories can be interpreted in several ways shows that nature is much richer than we have imagined.

Newton's laws of mechanics are a good example. They can be written in ways that are different from either the local, differential mode, or the global, action-at-a distance mode. It turns out that certain quantities involving kinetic and potential energy must be either a maximum or a minimum, and this condition can be used to calculate the path of any object subject to forces.[90] So we can say either that nature consists of fields interacting locally, or that it acts over large distances with nothing between them, or that it works by requiring certain energies to be either a maximum or a minimum. Which of these represents what is *really* going on? A reluctance to make a decision is behind the idea that only the mathematics matters. After all, each of these methods can be transformed into the others by doing some math, so the physical interpretation is not really important. This is a defensible position, but it is more useful and satisfying to maintain that *all* of these methods have their expressions in the natural world. All of them reveal an aspect of physical reality and all of them are useful for calculating and learning about new phenomena.

Any analysis of scientific truth must acknowledge the role of mathematics. The central position of mathematics in so much of science has been a source of delight and wonder to the mathematically inclined, as well as a barrier to those who have no mathematical training. Why is mathematics so important in science and what does it really tell us about physical reality? Arithmetic is a good place to start because all of analysis can be reduced to, and derived from, arithmetic.[91] The essence of this kind of mathematics is that it arises from the concept of numbers as successions and introduces the properties of "greater than" and "less than". Once these are defined, all else follows. The connection with the physical world starts at this basic level with the ideas of distance and of time. We can compare lengths and see that some are greater than others, and we can compare time intervals, some of which are greater than others. For any physical property that can exist in different amounts, the amounts can be ordered in the same

[90] These are Hamilton's principle and the principle of least action.

[91] Analysis starts with arithmetic and includes algebra, calculus, and their offshoots and generalizations. It is distinguished from geometry, which can be developed independently from the concept of number, and different from such disciplines as topology, which are pure logical structures independent of the idea of quantity.

way as numbers. Temperature, mass, pressure, and velocity, for example, all have this characteristic, so that measurements of each of them can be put into correspondence with a set of increasing numbers. Once this is done, these quantitative properties can be subject to all of mathematical analysis. Because the basic properties can be put into correspondence with numbers, we expect that the more complex results of applying mathematics to physics can be put into correspondence with something in nature.

Here is an example: In working out the mathematical description of a body falling to the ground, we recognize that the distance of the object from the ground is changing, and we construct a scale to measure the distance, thereby introducing the fundamental property of numbers. We also recognize that the time is changing, so we construct clocks that give us a numerical measure of time. The Galileo–Newton study of motion led to Newton's second law, thereby introducing the concept of mass and force, which can be assigned quantitative measures because time and distance were numerically quantified. Taking Newton a bit further, we find a certain quantity that is large when the object is high above the ground and low when it has competed its fall. At the same time, we see another quantity that is small when the object is moving slowly and large when it is falling rapidly. We notice that *the sum of these two quantities is always the same.* The first quantity is, of course, the potential energy, and the second is the kinetic energy. We have just described one instance of the law of the conservation of energy. This is a mathematical expression, but we assume that it has a physical counterpart. That is, we assume that energy exists in the natural world and that it is conserved.

It is important to note that there is a structure here. Time, distance, mass, force, velocity, acceleration, and energy are all related in a certain specific way. The structure is the result of physical research, and the mathematics expresses it in a concise, transparent shorthand from which the consequences of that structure can be found. That is why scientists often say that a certain equation is beautiful or that a mathematical theory, such as relativity is beautiful. The mathematics displays the inner structure of nature in a compelling manner whose range and power is immediately evident, all because measurements can be ordered in a way analogous to the ordering of numbers.

Of course, not all equations or theories are beautiful. Some of them are ugly and overly complicated, and these are usually suspect, and great efforts are expended to make them simpler, to test their validity, or lack thereof. The most successful theories—relativity, quantum mechanics, continuum mechanics, electrodynamics, thermodynamics, statistical mechanics—are all expressed in simple elegant equations based on very simple ideas. The equations codify an enormous amount of information in a line or two of mathematics. Their simplicity, their range, and their power make them beautiful.

24

The meaning of why

We are always searching for reasons. The perception of chains of causality is a major tool for coping with the everyday events that define our existence. The coffee spilled out of my cup because I banged it with my elbow; it burned my hand because it was hot; it was hot because it was brewed using the heating element in the coffee maker; the coffee maker was turned on because I threw a switch ... Without consciously thinking about it, I know that whatever happens arises from a series of causes, each of them having an answer to the question "why"? The causal mode of thought is essential because it teaches me to avoid actions that are harmful and to pursue those that are of benefit. In my simple example, I try not to knock over coffee cups because I do not like the result. The sense of causality goes deeper than conscious logic, as shown by the fact that all animals will try to avoid those conditions which bring them danger and pain, and seek those which will provide for their needs.

For us, the desire for causality goes far beyond the requirements of everyday life; it is the root of the ageless, universal urge towards religion and mysticism, and it is a major drive for the development of science.

It is the nature of causality in the scientific analysis of the physical world that concerns us here. It is closely related to the meaning of explanations for physical events and, indeed, to the question of whether or not explanations can even be found. In the sequence of events leading to my coffee spill, I was able to assign causes that depended on my own actions, and this gave a satisfying meaning to the question "why"? But the falling of the coffee to the floor is a different matter. It just fell and would have fallen even if I had spilled it on purpose, or if some other cause than my elbow had knocked over the cup. If someone else had knocked it, if a bird had flown into it, if a child's ball had hit it, or if an earthquake had shaken it, the result would be the same. Once the cup tipped over, the coffee would fall to the ground.

The question then is "why do things fall"? When the Greeks presumed that an object fell because its natural place was the Earth, and it was just trying to get there, they were giving a teleological answer to this question, as if the objects had a will that urged them home. In this way, they could understand why things fall. They did not recognize that, scientifically, the answer was really a tautology, amounting to saying that objects fall because they fall. The seventeenth-century natural philosophers looked for something deeper, and their experiments, observations, and logic culminated in Newton's law of universal gravitation. For many, the issue was settled. Objects fell to Earth because they were attracted to it by Newton's law. At the same time, they found the reason for the existence and properties of planetary orbits, for ocean tides, for the working of machines, and for a host of other mechanical phenomena. The ramifications of the Newtonian laws of mechanics and gravitation were so impressive and far-reaching that they were taken to be explanations. Newton's own position on the truth about gravitation was different. He did not believe that his laws explained anything. He explicitly stated that he was just working out their consequences and that a true explanation had to await further work. His law of universal gravitation, and its results, constitutes a *description* of how gravity works, not an explanation of *why* it exists.

Then along came Einstein, and once more, many believed that the great "why"? was at last answered. According to general relativity, gravity existed because the presence of matter produced a curvature in space, and moving objects followed this curvature. The law of gravitation has its origin in the way matter curves space. But is it not obvious that this simply replaces one question by another? Instead of asking why does the force of gravitation exist, we now ask why does matter curve space! So we go back to the theory and search it to find out that matter must curve space because only then are the laws of physics the same always and everywhere. The result of such universality is that space must be curved, that objects must move along these curves, and this gives us gravity.

We have come far. Gravity exists because space is curved and space is curved because physical laws are everywhere the same. But again, we have only pushed the need for explanation back one step. Let's grant that indeed the laws of physics are the same in any physically conceivable coordinate system, because observation and experiment show that the results of this assumption, as embodied in general relativity, are correct. The next question is "why should this be so"? There is no answer. The universality of physical laws is more than a philosophically satisfying concept, because its consequences are verified by experiment. But it is not an explanation in the same sense that my turning on a switch to heat my coffee is an explanation. In the physical world, the chain of causality is continually pursued to a point where there is nothing further, a fact that has been recognized for centuries and which resulted in the "First Cause" and "Prime Mover" arguments for the existence of God. The answer to the ultimate "why" is then easy: "because God wills it".

Compare the two explanations "turning on the switch" and "God is the Prime Mover" and see how similar they are. The quest for a "why" seems satisfied when we get to an intelligence (or an Intelligent Being) that takes some action or makes a decision. For most people the search stops right there. An independent free will has made a choice; no other explanation is needed. For others, this is not enough, and they want to know why that "free will" made that particular decision. We are then in the arena of the age-old battle between free will and determinism. Science has nothing useful to say about this.

For the physical sciences, the meaning of "explanation" and answers to the question "why" depend on what we are willing to accept. A commonly held view is that physical laws certainly are explanatory. Newton's law of universal gravitation was explanatory in that it stripped many unknowns of their mystery, uniting them into a coherent framework. General relativity went further, exposing the connection between observed gravity and the fundamental nature of space and time, while showing that the world followed an invariance that was not merely philosophical. For many, this is explanation enough. For many others, it is not. No matter how advanced or detailed scientific theory becomes, they will always have a "why" question of the sort that science cannot answer. The Galileo affair starkly exposed the difference between science and the desire for ultimate causality. The position of the Church was simple. God is the Cause of all things and if science contradicts God's word, then science is wrong. But science has nothing to say about God and ultimate causality. God may or not be an Ultimate Cause. Science has nothing to say about that question. God may or may not be acting continually to keep the universe working; He may or may not have set up initial laws and then just let them evolve; and God may or may not even exist! Science has no methods, no observations, no theories, no tools—no means whatever, to address these issues. It deals only with observable, objective facts and the logical inferences that can be made from them. Science can confront religion at only one point. If an interpretation of God's word is contradicted by a scientifically established fact, then that interpretation is wrong.

A famous example is that of the age of the Earth. In 1650–54, James Usher, a prominent Irish Protestant bishop, published a calculation of the age of the Earth. From a close analysis of the Bible, including the counting of years for generations of biblical figures, he concluded that the world was created on the evening of October 23, 4004 BC We must admire his precision. But Usher was not the first to make such calculations, and they were not restricted to Protestants. On a visit to Rome in 1587, Galileo learned of some Jesuit calculations in which a study of the Bible showed that the universe was created in 4160 BC Given the difficulties of counting biblical generations and epochs, the agreement between the two numbers is quite good.

The scientific evidence is overwhelming that these dates are wrong by many billions of years, so any rational person should reject them. But the Usher number figured importantly in the 1925 Scopes trial, and is still held to be true today by creationists who reject Darwinian evolution and instead assign the current existence of *everything* just as it is to God's direct action. So evolution must be wrong; life comes directly from God, not from random selection of "the fittest". It is an effect of "creation" or "intelligent design". Evolution claims to take place in bits and pieces, over enormous time scales, strongly affected by accidental and random factors, with an incredible amount of waste: so many blind alleys and so much suffering. And humanity is just another species in this ugly history, different only by virtue of a large brain, just as an elephant is different because of its large body, but otherwise just another animal.

All this is contrary to the idea that a personal God cares about people. So evolution is rejected. There must be some grand plan that includes human beings in some special way. There must be a God, directly connected to human beings, Who designed and created the universe with specific purposes in mind. The creationists feel so strongly about this that they spend great efforts and huge sums of money on attempts to have creationism (or its euphemism, "intelligent design") accepted as science and taught in schools as an alternative to evolution.

The Church that tried and convicted Galileo and the fundamentalists that deny Darwin are victims of the same mistake. The Aristotelian physics and the Ptolemaic astronomy embraced by the Church were wrong and they were proven to be wrong. And the creationists are wrong when they accept the Usher calculation or reject evolution. Both relied on authority, both were shown to be wrong by logical analysis of data and observation, and both thought that science was attacking religion. But neither the Copernican system nor evolution has anything to say about the truly important religious questions. The Church was seen to be ridiculous by sticking to Ptolemy, just as the creationists are regarded as mindless zealots by rejecting modern science. Such conflicts damage both science and religion, each of which needs to stay in its own domain. In the long run, religion suffers more, because many who claim to be religious promote unreal interpretations of nature. The credibility of religious thought is then damaged.

Science only provides descriptions of the physical world and has nothing to do with fundamental explanations, which are properly the province of epistemology and philosophy. Thus, when one billiard ball strikes another and sets it in motion, regarding the motion of one ball as being caused by the other looks right, but it is not appropriate, or even fruitful. We are stationary relative to the pool table, so when we see a stationary ball bounce away after being hit, it is natural to say that the first moving ball caused the other one to move. But if we choose a coordinate system that is stationary relative to the first ball, then it looks like it was hit by the second ball. Then we can say that the motion of the second (initially stationary) is caused by the first ball striking it.

Physics describes *what* happens when the two balls collide, not *why* it happens. Similarly, none of the theories of gravity contain an explanation; they just describe gravity. More advanced theories give better and more detailed descriptions, but nothing that can be called an explanation.

The only acceptable scientific answer to the question of the ultimate origin of physical phenomena is that this is just how the world is. What we experience is the result of the fundamental structure of nature. This is again a tautology, meaning that things are that way because they are. In science, the question "why is nature as it is"? should be replaced by the question "what is the basic structure of nature"? Here lies the sharp difference between the ancient physics of Aristotle and that of the moderns. Both examined the physical world and constructed descriptions of what was happening and how it happened, but the ancients were looking for ultimate causes. Of course, some moderns thought they were looking for final answers, but their work shows that they were really finding a description of the underlying physical reality behind the results of their observations and experiments.

The action-at-a-distance theorists concluded that nature was such that bodies could influence each other without any intervening medium. Proponents of the existence of the ether thought this was absurd; *something* had to be in those spaces to carry the forces. Descartes's vortices, Faraday's lines of force, Maxwell's fields, and even Einstein's curved space were alternatives to the unacceptable idea that physical actions could exist, and make themselves felt, in totally empty space. All concepts of the ether, from the ancient *plenum*, to the modern quantum fields, are attempts to describe the reality of nature at a deeper level than that of mere observations, experiments, and mathematics. We want the truth. Is there any hope of getting it, and what does truth in science really mean anyway?

Science is the method of learning about nature, so we have to start with the objective facts about nature. A systematic methodology has evolved to ascertain what the true facts really are. The methodology assures us that the facts are public, available to all, and not just the product of an individual mind, which may be subject to bias, incompetence, or hallucination. Some facts are absolute and others are not. When a particular object is dropped, it falls to the ground. Anyone looking at it can see, and everyone will agree, that it falls to the ground. This is an absolute fact. After seeing many objects fall to the ground, and none rise to the sky, we generalize this by stating that *all* objects will fall to the ground if they are let go. This is not an absolute fact, because the dropping of *all* existing bodies cannot be observed. It is a generalization of so many observations that we accept it as an absolute truth. But we know it is not truly absolute! If we drop a helium balloon, it will go up, not down. If we drop a bowling ball on a trampoline, it will go down, but then bounce up. If we attach a rocket motor to a cylinder, start it, and then let go, it will go up. It is easy to smile at these exceptions and remark that any idiot knows that things fall when there is no other force than the Earth's gravity acting on them and that is what the generalization really says. Yet it is important to see that generalizations are correct only if the conditions of their existence are specified. The full statement of any generalization must be sufficiently detailed that when exceptions are found, the reasons for them will be apparent. And even after all appropriate conditions are defined, and even after a great many cases are observed for which the generalization holds, this does not make it absolute, because it is not possible to test *every* possible case. But the greater the number of cases verified, the greater is our belief that the generalization is correct.

This was not a trivial issue in the history of gravitation. Let's recall the classic experiments on the speed of falling bodies. Before Galileo, it was commonly believed that heavy bodies fell more rapidly than light ones, and there was strong observational support for that belief. It was easy to see that lighter objects did fall more slowly than heavy ones. Drop a feather and a book and note which one hits the ground first. It was not until the role of air resistance was fully appreciated that a correct statement of the facts of falling bodies could be formulated. The rule of falling bodies that made future scientific advance possible is: "All bodies will reach the ground at the same time, if they are dropped from the same height, in a vacuum, at the same place on the Earth". The physics of motion and gravity was held back for centuries because of the failure to recognize the conditions under which the experiments were made.

Here is a rule: To generalize from a set of facts, make sure you have the facts, all the facts, and nothing but the facts.

Quantitative experimental data are essential. Knowing that all planets revolve around the Sun is an important observation, but no further advance could be made until Tycho Brahe measured the actual positions of their orbits and Kepler put them in simple mathematical form.

And measurements that are not a count of discrete entities are always approximate. They are different than the mere noting of an event, and can never give an absolute value. Let's consider a very simple measurement: that of the length between two scratches on a rectangular metal strip. Using a centimeter rule, we find that the measurement gives us 25 centimeters. Is this a truth? A look at the ruler shows that each centimeter is divided into tenths, so this is a rather good ruler. (We also see that the scratches are nearly one twentieth of a centimeter thick, so we take our measurement from the edges of the scratches.) We cannot measure more accurately than one tenth, so our measurement gives us that the length is between 24.9 and 25.1 centimeters. We do not know the "true length", only that it is between two limits. If it is important enough, we can increase the accuracy of the measurement by using better instruments, so, for example, we can find that the length is between 24.95 and 24.96 centimeters, but this still gives us a set of limits and not an absolutely true value for the length. All measurements have the same character in that they can only yield an approximation and never an exact number. The approximate nature of measurement has led some people (who do not understand the scientific process) to conclude that science cannot tell us *anything* absolute or true, that all scientific pronouncements are tentative and shaky, so science is not a valid way of finding truth. But it *is* an absolute fact that the distance between our marks is between 24.9 and 25.1 centimeters. And when the velocity of light is measured thousands of times using ever more precise instruments, and the experimental error gets ever smaller as the instruments get better, only the most stubborn skeptic would refuse to accept the constancy of the speed of life as a fact. Of course scientists recognize that experimental error can never be completely eliminated, and of course we recognize that for certain measurements we face the limitations of quantum mechanics, and of course we recognize that *in principle* nothing is absolutely certain, But we are not dealing with a world "in principle"; we live in a real world in which the accuracy we possess is developed to the point that we can deal with that real world. Yet we strive for ever increasing accuracy because sometimes small deviations from the expected values can result in large changes in our basic understanding of the world. It took accurate measurements of the velocity of light to establish its constancy. And what if we find, by making much more accurate measurements, that the velocity of light in a vacuum is *not* constant? This could well produce a revolution in our understanding of nature comparable to that of relativity. So we keep looking, even when we truly believe that we will find no revolution.

It pays to notice that, when describing the ancillary conditions that affect how a body falls, we said that the reasons for deviations from the ideal case must be found. We often use the language of causality as shorthand, but it is important that this habit, which arises from our desire to know the unknowable "why", not be raised to philosophical status.

Science can only describe. The ultimate reasons for physical laws, for our existence, for the meaning of life, for the origin of the universe and the place of human beings in it cannot be found by scientific methods. Science tells us "what"? not "why"? The questions of ultimate reasons and of final causes that go beyond the methods of science are questions of philosophy and religion. The confusion between "what"? and "why"? continues to sow mischief by pitting science against religion. Scientific knowledge per se has nothing to do with religion, with the relation of man to God, with morality or virtue or sin. Science provides only one boundary condition, and it is important enough to repeat: any philosophical or religious statement that contradicts scientific fact cannot be right. And any religion or philosophy that contradicts well-established scientific theory is most likely wrong, at least at the point of contradiction. These two propositions were dramatically illustrated by the case of Galileo, who proclaimed the heliocentricity of the solar system in spite of religious belief to the contrary, and by Descartes, who developed a theory of nature from philosophical beliefs that was detailed, complete, and wrong.

Facts alone do not satisfy us. They do not bring us closer to knowing how the world works. They simply note what we have seen. We need more, so once the facts are in hand, we search for some way of tying them all together. Newton inherited the facts about falling bodies and about planetary motion and developed the theory of universal gravitation, with the inverse square law of attraction, to bring all the facts about gravity into one simple description. This is much better than just knowing Kepler's three laws and Galileo's observations. Newton's theory was *universal*. It worked everywhere, all the time. Of course, we do not absolutely know that it is universal, because it is not possible to measure gravitational attraction for *all* the interacting bodies in the universe. We believe it is universal and accept it as so because every time a measurement is made, Newton's law is verified. Note that if a single instance were found of two gravitating bodies in which Newton's law did not hold, the law would be rejected.

The history of gravitation vividly displays the crucial importance of accurate measurement. At first, Newton would not commit himself to the inverse square law, even though he felt it was true. When applied to a comparison of the gravitational attraction between the Earth and the Moon, to that on the surface of the Earth, he got the wrong answer because the accuracy with which an essential datum was known, the radius of the Earth, was too poor. Newton could not embrace the inverse square law until more accurate results were available. Later, accurate determinations of planetary orbits, and their deviation from existing Newtonian calculations, led to the discovery of the trans-Uranic planets. And it was sufficiently accurate measurements of the speed of light that showed its independence from the state of motion and thus led to special relativity, which led to the general relativity theory of gravity.

The evolution of gravitational theories is a typical case of how scientific theories change with time as new information is acquired. This is often cited as another proof that science can never yield truth, that anything is possible, and that non-scientific ways of learning about nature are valid. The intellectual confusion and misunderstanding of science attending this position is extreme.

25

Final comments

The study of gravitation displays one of the greatest intellectual achievements of the human mind. It is a story of strong personalities driven by a passionate desire that dominated their lives: the desire to understand the true nature of reality. Nature's secrets can be divined only through an intense commitment by "monster minds" that tower above the great mass as Everest towers over ordinary hills. Those that chased after gravity had such intellects, and yet they were ordinary people in many other respects and had to deal with many of the human problems in their own make-up and in the world around them. Kepler had a truly miserable childhood, and had to battle Tycho's suspicions and disapproval, as well as his own magical mystic tendencies; Galileo contended with a powerful religious hierarchy and a fully entrenched conservative intellectual establishment, and his combative personality always worsened his situation; Newton was bedeviled by his own crusty, suspicious nature and messianic ego; Einstein was beset by Nazi anti-Semitism and was continually on the move. Their lives were full of drama and conflicts, some internal and some with the outside world. And so their work was not merely the result of quiet contemplation and careful pursuit of science. It was a triumph over many obstacles.

The progress of our knowledge about gravitation is a strong illustration of how extraordinary people looking at ordinary, mundane facts discover that nature is wonderful and strange, unlike anything we would ordinarily believe. Everyone has seen things move and everyone has at least an intuitive idea of constant velocity. And everyone has seen objects fall and everyone knows that *anything* will drop to the ground if it isn't supported. These are simple, commonplace facts. But thinking about them, making some observations, and doing some experiments, all in a commonplace, very human way, leads to conclusions that are hard to imagine. Who could imagine curved space-time, or that objects move in a gravitational field simply because any change from inertial motion would require some other force? The simple notion that the laws of nature are the same everywhere, for everybody, that seems so right and so philosophically satisfying means that matter curves space-time and that this curvature is a universal force holding the entire universe together. All this follows from applying ordinary logic to ordinary things. Yet only giants could do it.

Additional reading

The following is a list of the books I found most useful or most interesting. They range from texts for those with a mathematical background to biographies and popular works for the "celebrated intelligent layman". Each reader will know which ones to choose.

Bergmann, Peter Gabriel, *Introduction to the Theory of Relativity* (New York: Prentice-Hall, 1942).

Bergmann, Peter G., *The Riddle of Gravitation* (New York: Scribner's, 1968; revised and updated edition, New York: Dower, 1992)

Berlinski, David, *Newton's Gift* (New York: Simon and Schuster, 2000)

Bohm, David, *The Special Theory of Relativity* (New York: Benjamin, 1965; London: Routledge, 1996)

Born, Max, *Einstein's Theory of Relativity* (London: Mehuen, 1924; revised and enlarged edition, New York: Dover, 1962)

Brian, Denis, *Einstein: A Life* (New York: Wiley, 1996)

Bridgman, P. W., *A Sophisticate's Primer of Relativity* (Middleton, CT: Wesleyan University Press, 2nd edition, New York: Dover, 2002)

Butterfield, Herbert, *The Origins of Modern Science* (New York: Macmillan, 1960)

Clark, Ronald W., *Einstein: The Life and Times* (New York: Houghton Mifflin, 1971)

Chandrasekhar, S. *An Introduction to the Study of Stellar Structure* (Chicago: University of Chicago Press, 1939; New York: Dover, 1958)

Cohen, I. Bernard, and Westfall, Richard S., *Newton: Texts, Backgrounds, Commentaries* (New York: Norton, 1995)

Cohen, Morris, and Drabkin, J. E., *A Source Book In Greek Science* (New York: McGraw-Hill, 1948)

Crombie, A. C., *The History of Science from Augustine to Galileo* (London: Heinemann, 1959; New York: Dover, 1995)

Dugas, René, *A History of Mechanics* (Neuchatel, Switzerland: Editions du Griffon, 1988; New York: Dover, 1995)

Einstein, Albert, *The Meaning of Relativity* (Princeton: Princeton University Press, 1950)

Einstein, Albert, and Infeld, Leopold, *The Evolution of Physics* (New York: Simon and Schuster, 1938; New Clarion edition, 1960)

Ferguson, Kitty, *Tycho and Kepler* (Walker, 2002)

Ferguson, Kitty, *Prisons of Light* (Cambridge: Cambridge University Press, 1996)

Fermi, Laura, and Bernardini, Gilberto, *Galileo and the Scientific Revolution* (New York: Basic Books, 1961; New York: Dover, 2003)

Friedman, Herbert, *The Astronomers' Universe* (New York: Norton, 1990)

Galileo Galilei, *Discoveries and Opinions*, translated by Stillman Drake, (New York: Anchor, 1957)

Galileo Galilei, *Dialogue Concerning the Two Chief World Systems*, translated by Stillman Drake (New York: Modern Library, 2001)

Galileo Galilei, *Dialogue Concerning Two New Sciences*, a translated by Henry Crew and Alfonso de Salvio (Amherst NY: Prometheus, 1991)

Gamow, George, *Gravity* (New York: Anchor, 1962; New York: Dover, 2002)

Gleick, James, *Isaac Newton* (New York: Pantheon, 2003)

Hall, Marie Boas, *The Scientific Renaissance, 1450–1630* (New York: Harper, 1962; New York: Dover, 1994)

Hall, A. Rupert, *From Galileo to Newton* (New York: Harper and Row, 1963; New York: Dover, 1981)

Hamilton, James, *A Life of Discovery: Michael Faraday, Giant of the Scientific Revolution*, (New York: Random House, 2002)

Hawking, Stephen, *A Brief History of Time* (New York: Bantam, 1998)

Hawking, Stephen, and Penrose Roger, *The Nature of Space and Time* (Princeton NJ: Princeton University Press, 1996)

Jammer, Max, *Concepts of Force*, (Cambridge MA: Harvard University Press, 1957; New York: Dover, 1999)

Jammer, Max, *Concepts of Mass in Classical and Modern Physics* (Cambridge MA: Harvard University Press, 1961; New York: Dover, 1997)

Landau, L. D. and Rumer, G. B., *What Is Relativity?* (New York: Dover, 2003; original Russian edition, Moscow, 1959)

Lorentz, Hendrik, Einstein, Albert, Minkowski, and Weyl, Hemann, *The Principle of Relativity: A Collection of Original Memoirs* (New York: Dover,—a 1923 translation from the original German collection of 1922)

Machamer, Peter (ed.), *The Cambridge Companion to Galileo*, (Cambridge: Cambridge University Press, 1998)

MacLachlan, James, *Galileo Galilei: First Physicist*, (New York: Oxford University Press, 1997)

Mahon, Basil, *The Man Who Changed Everything: The Life of James Clerk Maxwell* (Chichester: Wiley, 2003)

Maxwell, James Clerk, *Matter and Motion* (Society for Promoting Christian Knowledge, London 1920; New York: Dover, 1952, 1981)

Newton, Isaa, *Opticks* (London: Bell, 1931, based on the fourth edition, London: 1750; New York: Dover, 1952)

Newton, Isaac, *The Principia*, translated by I. Bernard Cohen and Anne Whitman, preceded by '*A Guide to Newton's Principia* by I. Bernhard Cohen (University of California Press Berkeley, Los Angeles, London, 1999)

Poincaré, Henri, *The Value of Science* (Modern Science Library, 2001, translations of Poincaré's, *Science and Hypothesis: The Value of Science and Science and Method* published in French in 1903, 1905, and 1908)

Reichenbach, Hans, *From Copernicus to Einstein* (Modern Library, 1942)

Reston, James, Jr., *Galileo: A Life* (New York: Harper Collins, 1994; reprint, Washington DC: Beard, 2000)

Rindler, Wolfgang, *Essential Relativity*, (New York: Springer, 1977)

Schwinger, Julian, *Einstein's Legacy* (New York: Scientific American Books, 1986; New York: Dover, 2002)

Simpson, Thomas K., *Maxwell on the Electromagnetic Field: a Guided Study* (New Brunswick NJ: Rutgers University Press, 2001)

Smith, James H., *Introduction to Special Relativity* (New York: Benjamin, 1963; New York: Dover, 1995)

Sobel, Dava, *Galileo's Daughter* (Walker, New York, 1999; New York: Penguin, 2000)

Thorne, Kip S., *Black Holes and Time Warps* (New York: Norton, 1994)

Tolman, Richard C., *Relativity, Thermodynamics and Cosmology* (Oxford: Oxford University Press, 1934; New York: Dover 1997)

White, Michael, *Isaac Newton: The Last Sorcerer* (Fourth Estate, Reading, MA: 1997; Perseus, 1999)

White, Michael, *Galileo Galilei: Inventor, Astronomer and Rebel* (Woodbridge CT: Blackbirch, 1999)

Whittaker, Edmund, *A History of the Theories of Aether and Electricity* (London: Nelson, 1951, 1953; New York: Dover, 1989)

Index